GEOMETRY AND INTEGRABILITY

Many integrable systems owe their origin to problems in geometry and all are perhaps best understood in a geometrical context. This is especially true today when the heroic early days of study of KdV-type integrability are over. The problems that can be solved using the inverse scattering transformation are now well studied and there are diminishing returns in this direction. Two major techniques have emerged more recently for dealing with multi-dimensional integrable systems: Twistor theory and the d-bar method, both of which form the subject of this book. It is intended to be an introduction, though by no means an elementary one, to current research on integrable systems in the framework of differential geometry and algebraic geometry.

This book arose from a semester, held at the Feza Gursey Institute, to introduce advanced graduate students to this area of research. The articles are all written by leading researchers and are designed to introduce the reader to contemporary research topics.

LONDON MATHEMATICAL SOCIETY LECTURE NOTE SERIES

Managing Editor: Professor N.J. Hitchin, Mathematical Institute,
University of Oxford, 24–29 St Giles, Oxford OX1 3LB, United Kingdom

The titles below are available from booksellers, or, in case of difficulty, from Cambridge
University Press at www.cambridge.org.

London Mathematical Society Lecture Note Series. 295

Geometry and Integrability

Edited by

Lionel Mason
University of Oxford

and

Yavuz Nutku
Feza Gurzey Institute, Instanbul

CAMBRIDGE
UNIVERSITY PRESS

CAMBRIDGE
UNIVERSITY PRESS

University Printing House, Cambridge CB2 8BS, United Kingdom

One Liberty Plaza, 20th Floor, New York, NY 10006, USA

477 Williamstown Road, Port Melbourne, VIC 3207, Australia

314-321, 3rd Floor, Plot 3, Splendor Forum, Jasola District Centre, New Delhi - 110025, India

103 Penang Road, #05-06/07, Visioncrest Commercial, Singapore 238467

Cambridge University Press is part of the University of Cambridge.

It furthers the University's mission by disseminating knowledge in the pursuit of
education, learning and research at the highest international levels of excellence.

www.cambridge.org
Information on this title: www.cambridge.org/9780521529990

First published 2003

A catalogue record for this publication is available from the British Library

Library of Congress Cataloging in Publication data
Geometry and integrability / edited by L. J. Mason, Y. Nutku.
p.   cm. – (London Mathematical Society lecture note series; v. 295)
Includes bibliographical references and index.
ISBN 0 521 52999 9
1. Global differential geometry.   2. Twistor theory.   3. Fiber spaces
(Mathematics) I. Mason, L. J. (Lionel J.)   II. Nutku, Yavuz.
III. London Mathematical Society lecture note series, 295.
QA670.G463   2003
516.3´62 – dc21   2003048456

ISBN   978-0-521-52999-0   Paperback

To our families

Contents

Contributors

F. Calogero
Università di Roma 'La Sapienza'

R. Y. Donagi
University of Pennsylvania

L. J. Mason
The Mathematical Institute, Oxford

P. M. Santini
Università di Roma 'La Sapienza'

K. P. Tod
The Mathematical Institute, Oxford

N. M. J. Woodhouse
The Mathematical Institute, Oxford

Preface

Integrable systems continue to fascinate because they are examples of systems with nontrivial nonlinearities that one can nevertheless systematically analyse and often solve exactly analytically. However, there is no royal road to complete integrability, or even a precise all-encompassing definition and so, instead, one must resort to patterns and themes. This volume is concerned with a theme that emerges time and again of the deep links that integrability has with geometry. The motivation for holding a research semester devoted to 'Geometry and Integrability' at the Feza Gürsey Institute was precisely for the purpose of exposing students and post-docs to modern geometrical structures that form the natural setting for completely integrable systems.

1
Introduction

Lionel Mason

The Mathematical Institute, 24-29 St Giles, Oxford OX1 3LB, UK

1.1 Background

Integrable systems are systems of partial or ordinary differential equations that combine nontrivial nonlinearity with unexpected tractability. Often one can find large families of exact solutions, and general methods for generic solutions. This volume is concerned with the deep links that integrability has with geometry. There are two rather different ways that geometry emerges in the study of integrable systems.

1.1.1 Geometrical context for integrable equations

The first is from the context of the differential equations themselves: even those integrable equations whose origins, perhaps in the theory of water waves or plasma physics, seem a long way from geometry can usually be expressed in the context of symplectic geometry as possibly infinite dimensional Hamiltonian systems with many conserved quantities and often with much more further structure. But geometry is itself also a rich source of integrable systems; one of the first examples of a completely integrable nonlinear partial differential equation, the sine-Gordon equation first appeared in the 19th century theory of surfaces, as a formulation of the constant mean curvature condition on a 2-surface embedded in Euclidean 3-space. Now there are many more examples from geometry in many dimensions, from the two-dimensional systems given by harmonic maps from Riemann surfaces to symmetric spaces, to the anti-self-duality equations in 4-dimensions and more generally quaternionic structures in $4k$-dimensions.

The contributions of Tod, Mason and Woodhouse focus on the anti-self-duality equations either on a Yang–Mills connection on a vector

1

bundle over $\mathbb{R}^4$, or on a 4-dimensional conformal structure. The systems discussed by Santini also have a geometric origin, in their discrete form as quadrilateral lattices, and in their continuous limits as conjugate nets. The reductions and specializations of these systems then form many more geometrical examples of integrable systems: although the systems discussed by Donagi are presented as arising from complex algebraic geometry rather than Riemannian geometry, they have their origin in reductions of the real anti-self-dual Yang-Mills equations.

1.1.2 Geometrical transforms and solution methods for differential equations

The second way that geometry appears in the theory of integrable systems is in the transforms and solution methods that are brought to bear on integrable systems. There are many different strands here. The symplectic framework for integrable equations leads to the first definition of an integrable system, that due to Arnol'd and Liouville, in terms the existence of sufficiently many constants of motion satisfying various requirements. The Arnol'd–Liouville theorem leads to a transform of the system to action-angle variables by quadratures in which the action variables are constant and the motion is linear in the angle variables. In fact many interesting integrable systems admit further structures that imply Arnol'd–Liouville integrability. Those considered by Donagi are algebraically completely integrable so that the structures in question are complexified and required furthermore to be algebraic. Another structure that guarantees complete integrability is a bi-Hamiltonian structure.

These structures in finite dimensions lead, at least in principle, to the general solution by quadratures. Integrable partial differential equations can often be expressed as infinite dimensional examples of systems satisfying the Arnol'd–Liouville requirements often by virtue of admitting a bi-Hamiltonian structure. However, the infinite number of degrees of freedom mean that one can no longer solve the system in a finite number of quadratures. Nevertheless, new techniques become available. On the one hand there are hidden symmetries, both discrete, such as Backlund transforms, and continuous, such as those generated by flows associated to the Arnol'd–Liouville constants of motion, and these can help generate new exact solutions. But also there are transforms that apply to general solutions; historically, the inverse scattering transform was the first important example of this and was used to provide the transform to action angle variables for solutions subject to rapidly decreasing boundary conditions in precise analogy with the transform provided by the finite dimensional Arnol'd–Liouville theorem.

There are now a number of such transforms such as the inverse spectral transform, the Penrose and Ward transforms and so on. A remarkable feature of many of these transforms is the appearance of sophisticated complex holomorphic, and often even algebraic geometry. This complex analysis often plays a deep role in the finite dimensional case also. In the contribution of Woodhouse we see twistor theory as providing a similar transform between solutions to integrable equations and geometric structures, holomorphic vector bundles, that can be described in terms of free functions. This construction has the additional benefit that it applies to the general local analytic solution. A related method is based on the non-local $\bar{\partial}$-problem, so called $\bar{\partial}$-dressing. In the local case this is often simply an independent formulation of the twistor correspondence, but in the non-local case, such constructions go beyond standard twistor theory.

1.2 The contributions

The following is intended to provide some introduction to, and context for, the various contributions. I should make a disclaimer here that the context and background I give are perhaps rather one-sided and reflect my own point of view; there are a number of different points of view that might be taken on this material that are not presented here!

1.2.1 Notes on reductions of the anti-self-dual Yang-Mills equations and integrable systems, L. J. Mason; Curvature and integrability for Bianchi-type IX metrics, K. P. Tod; Twistor theory and integrability, N. M. J. Woodhouse

These contributions are connected by an overview on the theory of integrable systems based on reductions of the anti-self-dual Yang-Mills (ASDYM) equations and anti-self-dual conformal structures.

The ASDYM equations can be thought of as integrable by virtue of the existence of the Ward correspondence between solutions to these equations and holomorphic vector bundles on an auxilliary complex manifold, twistor space. For ASDYM fields on Minkowski space, twistor space is a portion of $\mathbb{CP}^3$, complex projective 3-space. If one allows the transform between solutions to the ASDYM equations and twistor data, this construction amounts to providing, in a geometric form, the general solution to the ASDYM equations. There is a similar construction due to Penrose giving a correspondence between anti-self-dual conformal structures and deformations of the complex structure on twistor space.

A key observation of Richard Ward's is that many of the most famous integrable equations are symmetry reductions of the ASDYM equations. The various aspects of the integrability of such reductions of the ASDYM equations can then be understood by reduction of the corresponding theory for the full ASDYM equations.

The contribution of Mason concerns the integrability of the ASDYM equations that can be understood without using twistor theory. Thus its Lax pair, Backlund transformations, Hamiltonian formulation and recursion operator and hierarchy are presented. Some of the more significant reductions are reviewed also.

Paul Tod's lectures on spinor calculus and conformal invariance were taken from his book with Huggett, Introduction to Twistor Theory (second edition), published by CUP as LMS Student-Text 5, and so are not included here. The book gives useful details of space-time geometry that provide a background for the twistor correspondence and the interested reader can refer to it for full details.

Nick Woodhouse's contribution is an introduction to twistor methods and explains how the Ward transform applies to ASDYM fields and descends to provide correspondences for reductions of ASDYM fields. In particular it is shown how twistor methods can give new insight into the KdV equations and the isomonodromy problem that arises in the study of Painlevé equations. One aspect of integrability that emerges particularly clearly is a 'geometric' explanation of the Painlevé test for integrability. The lectures build on those of Tod and Mason.

There is a further contribution from Paul Tod which concerns various equations on metrics in 4-dimensions that admit an $SU(2)$ symmetry. The metric may be required to be Kahler, Einstein or have anti-self-dual Weyl tensor. The latter equation is usually thought to imply integrability because of Penrose's twistor correspondence. With this symmetry, the equations reduce to ODE's. If the metric is Einstein, it is no restriction to assume it is diagonal (although it is a nontrivial restriction for general anti-self-dual conformal structures). When Ricci flat, one obtains (with a further assumption) the Chazy equation. This is somewhat of a novelty for integrable systems theory as this equation admits solutions with movable natural boundaries, contradicting the Painlevé property. An explanation of this paradox is proposed.

1.2.2 Geometry and integrability, R.Y. Donagi

The contribution of Donagi is concerned with the theorem that the Moduli space of 'meromorphic Higgs bundles' over a Riemann surface Σ has the structure of an algebraically completely integrable system. This combines the symplectic geometry underlying the Arnol'd–Liouville definition of an integrable system with algebraic geometry. The Arnol'd–Liouville definition of a completely integrable system as above can be abstracted by taking an integrable system to mean a Poisson manifold, M, with sufficiently many commuting Hamiltonians, the collection being thought of as a map from $H : M \mapsto \mathbb{R}^n$, satisfying certain technical requirements to guarantee satisfactory global properties. This definition can be complexified so that M is a complex manifold and the Poisson structure is a complex holomorphic bivector and $H : M \mapsto \mathbb{C}^n$ are holomorphic. The algebraic condition is then that M be an algebraic manifold with all the structures being expressible in terms of algebraic functions of algebraic coordinates on M. Although this definition might seem somewhat special, there are a remarkable number of interesting systems that turn out to be integrable in this way.

A Higgs bundle E is a holomorphic vector bundle equipped with a global holomorphic section, the Higgs field Φ, of the associated bundle of 1-forms with values in the endomorphisms of E, $\mathrm{End}(E) \otimes \Omega^1(\Sigma)$. These first arose in the context of Hitchin's study of reductions of the anti-self-dual Yang-Mills equations on a connection on a bundle over Euclidean $\mathbb{R}^4$ by two translational symmetries. Remarkably, the reduced sytem acquires 2-dimensional conformal invariance and so makes sense on an arbitrary Riemann surface. The anti-self-duality condition reduces to equations on a connection and Higgs field on the Riemann surface; the Higgs field should be holomorphic and the curvature of the connection be given in terms of the Higgs fields. According to the philosophy of the contributions by Woodhouse, Tod and Mason, this Hitchin system is an integrable system. Since it is a system of elliptic partial differential equations, it doesn't naturally fall into a Hamiltonian framework. However, the space of solutions on a compact Riemann surface is finite dimensional and one might expect this moduli space to inherit some vestige of integrability.

Hitchin proves that, for a compact Riemann surface and certain bundles, a solution is determined just by the holomorphic data of the holomorphic vector bundle and Higgs field. Thus the study of the moduli space can be reduced to a problem in complex geometry and this is the approach that is adopted in this article. Naively the moduli space can be thought of as the cotangent bundle of the moduli space of holomorphic vector bundles on Σ as the Higgs fields are Serre-dual to deformations of the complex structure on a holomorphic vector bundle. Thus the Higgs bundle moduli space is a complex phase space. Furthermore, the coefficients of the characteristic polynomial of the Higgs field can be thought of as defining a system of commuting Hamiltonians and so one has a complex (holomorphic) integrable system which turns out to be algebraic. However, there are a number of technicalities concerning stability and semi-stability that need to be addressed to make these ideas precise, and render the above discussion heuristic.

In keeping with the expository aim of the lectures, the bulk of these notes concern not the theorem and its applications, but the many ingredients which go into its proof. Students with a fairly modest background in geometry should be able to work through these notes, learning a fair amount of algebraic geometry and symplectic geometry along the way, and may be motivated to follow some of the leads in the last section towards open problems and further development of the subject.

1.2.3 The $\bar{\partial}$ dressing method and integrable geometries, P. Santini

In the previous contributions, it can be seen that a prominent role is played by complex structures. One way of formulating a complex structure is in the form of a $\bar{\partial}$-operator and, in the case of the Ward transform, the inverse transform from twistor data to the solution on space-time requires the solution of a linear $\bar{\partial}$-equation. Dressing can be understood as a process by which one takes the transform for a well understood, perhaps trivial, solution where all the ingredients of the tansform are known, and then change the $\bar{\partial}$-data that appears in the $\bar{\partial}$-equation to give a more general solution (perhaps the general solution). Over the last few decades such methods have been developed (independently of twistor theory) and extended to include a non-local element in the $\bar{\partial}$-equation, so that the source term in the $\bar{\partial}$-equation is given by integrating against a kernel. These non-local terms seem to be essential for certain systems in 2+1 dimensions such as the KP equations etc..

In this contribution the $\bar{\partial}$-dressing method is shown to apply to certain integrable geometric structures: quadrilateral lattices, a discrete system consisting of lattices in which each elementary quadrilateral is planar, and its continuous limit, the conjugate net, a system studied by Darboux.

The connection between the $\bar{\partial}$-dressing method and these integrable geometries relies upon the following facts:

(1) the simple, linear dependence of the $\bar{\partial}$ data on the coordinates, described by the given linear differential and/or difference equations, defines some basic elementary singularities in the complex plane of the spectral parameter λ (the complex parameter with respect to which the $\bar{\partial}$-problem is defined): essential singularities, poles and branch points, in which the coordinates appear as parameters of the essential singularities, positions of the poles and strength of the branch points.

(2) These elementary singularities and their defining equations have often an elementary and basic geometric meaning. For instance, (a) the matrix equation $\psi_{0x} = i\lambda\sigma_3\psi_0$ and its solution $\psi_0(x,\lambda) = \exp(i\lambda x\sigma_3)$ define the Frenet frame of a straight line in $\mathcal{R}^3$, parallel to the third axis with constant torsion λ and arclength x; (b) the vector difference equations: $\Delta_i\psi_{0j} = 0$, $i = 1,..,N$, $j = 1,..,M$ define the tangent vectors $\psi_{0j} = (0,..,\lambda^{\theta_j},..,0)^T$ of an N - dimensional regular lattice in $\mathcal{R}^M$.

(3) Through the $\bar{\partial}$ dressing method the above basic elementary functions ψ_0 get dressed into new functions ψ which satisfy dressed linear equations in configuration space, whose integrability conditions are the integrable nonlinear systems. In this dressing procedure, the original geometric meaning is usually preserved and suitably deformed. For instance, the linear equation of example (a) is dressed up into $\psi_x = (i\lambda\sigma_3 + Q)\psi$ and describes an arbitrary curve in $\mathcal{R}^3$; while the linear equations of example (b) are dressed up into the linear equations $\Delta_i\psi_j = q_{ji}\psi_i$, $i = 1,..,N$, $j = 1,..,M$ which describe the *planarity* of the elementary quadrilaterals of the N-dimensional lattice (what we call: *a quadrilateral, or planar lattice*).

(4) The associated $\bar{\partial}$ problem provides at the same time:

(i) large classes of solutions of the above geometries, which can therefore be called "integrable";

(ii) geometrically distinguished symmetry transformations and symmetry reductions of the above geometries.

1.2.4 Differential equations featuring many periodic solutions, F. Calogero

The contribution of Francesco Calogero, who is the originator of modern super-integrable systems amongst many other things, shows a way to obtain evolutionary PDEs which possess many periodic solutions. This development has obvious potential in the context of applications (especially in the modelling of periodic phenomena), but it also sheds light (as more fully shown in other papers by Calogero and others) on a rather fundamental question: the connection between the integrability of evolution equations and the analyticity in complex time of the solutions of such equations, an issue related to the 'Painlevé property'.

1.3 Conclusion

There are many areas of interaction between geometry and integrability that have not been touched on here — the infinite-dimensional grassmanians of Segal & Wilson, the theory of quaternion-Kahler manifolds, the various special integrable classes of two-surfaces embedded into symmetric spaces and so on, but it is to be hoped that these articles will stimulate the reader into further study.

Acknowledgements: I am grateful to Professor Nutku and TUBITAK for the invitation to the Feza Gürsey institute and for their generous hospitality. I should also like to thank Professor Nutku and the contributors for helpful paragraphs in writing this introduction.

2

Differential equations featuring many periodic solutions

F. Calogero

Dipartimento di Fisica, Universit di Roma "La Sapienza",
Istituto Nazionale di Fisica Nucleare, Sezione di Roma

francesco.calogero@uniroma1.it, francesco.calogero@roma1.infn.it

Mathematics Subject Classification 2000: 34C25, 35B10
Physics and Astronomy Classification Scheme: 02.30.Hq, 02.30Jr

Abstract

A simple trick is reviewed, which yields differential equations (both ODEs and PDEs) of evolution type featuring lots of periodic solutions. Several examples (PDEs) are exhibited.

2.1 Introduction

Recently a simple trick has been introduced that allows us to manufacture evolution equations (both ODEs and PDEs) which possess lots of periodic solutions – in particular, *completely periodic* solutions corresponding, in the context of the initial-value problem, to an *open* set of initial data of *nonvanishing* measure in the space of initial data [1]-[5]. The purpose and scope of this presentation is to review this trick – most completely introduced and described in [5] – and to display, and tersely discuss, certain new (classes of) evolution PDEs yielded by it; the alert reader, after having grasped the main idea, can easily manufacture many more examples, possibly also featuring several dependent and independent variables – here for simplicity we restrict attention to just *one (complex)* dependent variable and to just *two (real)* independent variables (the standard $1 + 1$ case: one 'time' and one 'space' variables only).

The trick is described tersely in Section 2.2. Some examples of evolution equations – different from those reported in [5] – are displayed

9

in Section 2.3, which should be immediately seen by the browser who wishes to decide whether to invest time in reading the rest of this paper. Justification for these examples – namely, arguments that justify the expectation that these evolution equations indeed feature lots of periodic solutions – are given in Section 2.4, and in some cases they are backed by the display there of some periodic solutions.

2.2 The trick

Suppose that the function φ of the two *complex* variables ξ, τ, $\varphi \equiv \varphi(\xi, \tau)$, satisfies an evolution equation in the (time-like) variable τ, and that the structure of this evolution equation guarantees that there exist a lot of solutions $\varphi(\xi, \tau)$ which are holomorphic in τ in an open disk of radius $1/\omega$ centered at $\tau = i/\omega$ in the complex τ-plane (where ω is a *positive* constant), and that are as well holomorphic in ξ in an open disk of radius ρ (where ρ is another *positive* constant, possibly arbitrarily large) centered at $\xi = 0$ in the complex ξ-plane. Then introduce a (complex) function $w \equiv w(x, t)$ of the two *real* variables x, t by setting

$$w(x, t) = \exp(i\lambda\omega t)\varphi(\xi, \tau) \tag{2.1}$$

with

$$\tau = \left[\exp(i\omega t) - 1\right]/(i\omega), \tag{2.2}$$

so that

$$\dot{\tau} \equiv d\tau/dt = \exp(i\omega t), \tag{2.3}$$
$$\tau(0) = 0, \dot{\tau}(0) = 1, \tag{2.4}$$

and

$$\xi = x \exp(i\mu\omega t). \tag{2.5}$$

It is then clear that, if λ and μ are two *rational* numbers, all the nonsingular functions $w(x, t)$ defined by (2.1) are, at least for $|x| < \rho$, *completely periodic* functions of the *real* independent variable t, with a period which is an integer multiple of $2\pi/\omega$.

On the other hand, if $\varphi \equiv \varphi(\xi, \tau)$ is determined by the requirement to satisfy an evolution equation of *analytic* type, say

$$\varphi_\tau = F(\varphi, \varphi_\xi, \varphi\xi, xi, \ldots, \xi, \tau) \qquad (a) \tag{2.6}$$

or

$$\varphi_{\tau\tau} = F(\varphi, \varphi_\xi, \varphi_{\xi,\xi}, \ldots, \xi, \tau) \qquad (b) \qquad\qquad (2.7)$$

with F an *analytic* function of all its arguments, then it is indeed clear, from the standard existence/uniqueness/analyticity theorem for the initial value-problem of analytic evolution PDEs, that there exist a set of solutions, of nonvanishing measure in the functional space of all solutions, that satisfy the requirements specified above, namely are *holomorphic* in the variables and sectors specified above. This is clear if one imagines obtaining these solutions $\varphi \equiv \varphi(\xi, \tau)$ of (2.6), (2.7) by solving an initial-value problem, with the initial datum assigned at $\tau = 0$ and such that the right-hand side of (2.6), (2.7), evaluated at $\tau = 0$, is *sufficiently small* (and, in the case of (2.7), the additional condition that $\varphi_\tau(\xi, 0)$ also be *sufficiently small*, say for all values of the complex variable ξ such that $|\xi| < \rho$).

The trick consists now in inserting the *ansatz* (2.1) with (2.2), (2.3), (2.4) and (2.5) in the evolution equation, say of type (2.6) and (2.7), satisfied by $\varphi \equiv \varphi(\xi, \tau)$, and thereby to obtain an evolution equation for $w \equiv w(x, t)$ that clearly then has a lot of solutions *completely periodic* in t – at least in an appropriately restricted space region (say, $|x| < \rho$; we will not keep repeating this condition below, but the reader should not forget it). What makes this development *interesting* is the possibility that the evolution equations for $w \equiv w(x, t)$ so manufactured have a *neat structure* – a possibility already demonstrated elsewhere [1]-[5] and also displayed immediately below.

2.3 Evolution equations featuring lots of periodic solutions

In this section we display, with no comments other than those needed to explain the notation, examples of evolution PDEs which possess lots of periodic solutions – in the sense explained above. These equations are obtained via the trick described in the preceding Section 2.2, as demonstrated in the following Section 2.4. Two such equations (or rather, classes of such equations) read as follows:

$$w_t - i\Omega w = \sum_{\substack{p_0, p_1, p_2, \ldots = 0 \\ \sum\limits_{n=0} p_n = p}} a_{p_0 p_1 p_2 \ldots} (w)^{p_0} (w_x)^{p_1} (w_{xx})^{p_2 \cdots}, \qquad (2.8)$$

$$w_{tt} - i[(p+3)/2]\Omega w_1 - [(p+1)/2]\Omega^2 w =$$

$$\sum_{\substack{p_0,p_1,p_2,\ldots=0 \\ \sum_{n=0} p_n=p}}^{p} a_{p_0 p_1 p_2 \ldots}(w)^{p_0}(w_x)^{p_1}(w_{xx})^{p_2}\cdots. \quad (2.9)$$

Here, and below, $w \equiv w(x,t)$ is the (complex) dependent variable, x and t are the ('space' and 'time', hence *real*) independent variables, subscripted independent variables denote (partial) derivatives, Ω is an arbitrary *real* and *nonvanishing* (hereafter, without loss of generality, *positive*, $\Omega > 0$) constant, p is an arbitrary *positive integer* ($p > 1$ so that these evolution equations are indeed nonlinear), the summation indices p_n are of course *nonnegative integers* not larger than p, and the constants $a_{p_0 p_1 p_2 \ldots}$ are *arbitrary* (possibly complex). These two evolution PDEs, of *first-order*, respectively *second-order*, have lots of solutions which are completely periodic with periods $T = 2\pi/\Omega$, respectively $T = 2\pi/\Omega$ (if p is odd) or $T = 4\pi/\Omega$ (if p is even).

Two other (classes of) evolution PDEs read as follows:

$$w_1 - i\lambda w w - i\mu\omega x w_x$$

$$= \sum_{\substack{p_0,p_1,p_2,\ldots=0 \\ \sum_{n=0} p_n=p, \ \sum_{n=1} np_n=P}}^{p} a_{p_0 p_1 p_2 \ldots}(w)^{p_0}(w_x)^{p_1}(w_{xx})^{p_2}\cdots$$

$$+ \sum_{\substack{q_0,q_1,q_2,\ldots=0 \\ \sum_{n=0} q_n=q, \ \sum_{n=1} nq_n=Q}}^{q} b_{q_0 q_1 q_2 \ldots}(w)^{q_0}(w_x)^{q_1}(w_{xx})^{q_2}\cdots$$

$$+ \sum_{j=0} \sum_{\substack{r_0^{(j)},r_1^{(j)},r_2^{(j)},\ldots=0 \\ \sum_{n=0} r_n^{(j)}=r^{(j)} \\ \sum_{n=1} nr_n^{(j)}=R^{(j)}}}^{r(j)} c^{(j)}_{r_0(j)r_1^{(j)}r_2^{(j)}\ldots}(w)^{r_0^{(j)}}(w_x)^{r_1^{(j)}}(w_{xx})^{r_2^{(j)}}\cdots \quad (2.10)$$

and

$$w_{tt} - 2i\mu\omega x w_{xt} - i(2\lambda+1)\omega w_t$$

$$= \mu^2\omega^2 x^2 w_{xx} + (\lambda+1)\mu\omega^2 x w_x + \lambda(\lambda+1)\omega^2 w$$

$$+ \sum_{\substack{p_0,p_1,p_2,\ldots=0 \\ \sum_{n=0} p_n=p, \ \sum_{n=1} np_n=P}}^{p} a_{p_0 p_1 p_2 \ldots}(w)^{p_0}(w_x)^{p_1}(w_{xx})^{p_2}\cdots$$

$$+ \sum_{\substack{q_0,q_1,q_2,\ldots=0 \\ \sum_{n=0} q_n=q, \ \sum_{n=1} nq_n=Q}}^{q} b_{q_0 q_1 q_2 \ldots} (w)^{q_0} (w_x)^{q_1} (w_{xx})^{q_2} \cdots$$

$$+ \sum_{j=0}^{r^{(j)}} \sum_{\substack{r_0^{(j)},r_1^{(j)},r_2^{(j)},\ldots=0 \\ \sum_{n=0} r_n^{(j)}=r^{(j)} \\ \sum_{n=1} nr_n^{(j)}=R^{(j)}}} c_{r_0^{(j)} r_1^{(j)} r_2^{(j)} \ldots}^{(j)} (w)^{r_0^{(j)}} (w_x)^{r_1^{(j)}} (w_{xx})^{r_2^{(j)}} \cdots . \tag{2.11}$$

In these evolution PDEs, (2.10) and (2.11), the arbitrary constant ω is real and *nonvanishing* (hence without loss of generality we hereafter assume it is *positive*, $\omega > 0$); the constants $a_{p_0,p_1,p_2,\ldots}, b_{q_0,q_1,q_2,\ldots}, c_{r_0^{(j)},r_1^{(j)},\ldots}^{(j)}$ are *arbitrary* (possibly complex); $p_n, q_n, j, r_n^{(j)}$ are *nonnegative integers* over which the sums run, subject to the constraints characterized by the parameters $p, P, q, Q, r^{(j)}$; the limit of the sum over j is determined by the limits over the integers $r^{(j)}$ and $R^{(j)}$, see below (this might entail this sum is altogether missing); the *positive integers* p, q and the *nonnegative integers* P, Q are arbitrary, except for the condition

$$(p-1)Q - (q-1)P \neq 0, \tag{2.12}$$

but we assume (without loss of generality) $p \neq q$, $p \neq r^{(j)}$, $q \neq r^{(j)}$ and $r^{(j)} \neq r^{(k)}$ if $j \neq k$, hence (again, without loss of generality) we set

$$p < q < r^{(0)} < r^{(1)} < r^{(2)} < \cdots . \tag{2.13}$$

The *nonnegative integers* $R^{(j)}$ are given by the formula

$$R^{(j)} = (p-q)^{-1} \left[pQ - qP + (P-Q)r^{(j)} \right], \tag{2.14}$$

and the corresponding positive integers $r^{(j)}$ are arbitrary except for the condition that $R^{(j)}$, as given by this formula, (2.14), be itself a *nonnegative integer* (note that the relations $r^{(j)} = \sum_{n=0} r_n^{(j)}, R^{(j)} = \sum_{n=1} nr_n^{(j)}$ also hold, see (2.10) and (2.11)). Finally the *rational numbers* λ and μ in (2.10) are given by the formulas

$$\lambda = (Q-P)/[(p-1)Q - (q-1)P], \tag{2.15}$$
$$\mu = (p-q)/[(p-1)Q - (q-1)P], \tag{2.16}$$

while the *rational numbers* λ and μ in (2.11) are given by the formulas

$$\lambda = 2(Q - P)/[(p - 1)Q - (q - 1)P], \tag{2.17}$$
$$\mu = 2(p - q)/[(p - 1)Q - (q - 1)P]; \tag{2.18}$$

note the consistency of these formulas, (2.15), (2.16) and (2.17), (2.18), with the condition (2.12).

These evolution PDEs, (2.10), (2.11) and (2.17), (2.18), have lots of solutions multiply periodic with the 3 periods $T_1 = 2\pi/\omega$, $T_2 = T_1/|\lambda|$ and $T_3 = T_1/|\mu|$ (of course with λ and μ given by (2.15), (2.16) respectively (2.17), (2.18), hence *completely periodic* (possibly only for a restricted set of values of the real "space" variable x , say $|x| < \rho$) with a period T that is the *smallest common integer multiple* of these 3 periods T_1, T_2, T_3. (In making this argument we implicitly assume that neither μ nor λ vanish; the first condition is indeed guaranteed by the condition (2.13), see (2.16) and (2.18); on the other hand λ could vanish, in which case the reference made above to the period T_2 should be ignored, namely in this case there would be a lot of solutions multiply periodic with the 2 periods T_1 and T_2, hence *completely periodic* with a period T that is the *smallest common integer multiple* of these 2 periods T_1, T_3.)

An example of an evolution PDE belonging to the class (2.8) (with $p = 4$) reads as follows:

$$w_1 - i\Omega w = a_4 w^4 + a_{13} w w_x^3 + a_{3001} w^3 w_{xxx} + a_{2110} w^2 w_x w_{xx}. \tag{2.19}$$

For $a_4 = a_{13} = 0$, $a_{2110} = 3a_{3001}$, or $a_4 = a_{13} = 0$, $a_{2110} = (3/2)a_{3001}$, this evolution PDE, (2.19), is C-integrable, while for $a_4 = a_{13} = a_{2110} = 0$ it is S-integrable. [6], [7]. This evolution PDE, (2.19), possesses of course many solutions *completely periodic* with period $2\pi/\Omega$.

An example of an evolution PDE belonging to the class (2.10) (with $p = 1$, $P = 3$, $q = 3$, $Q = 2$, $r^{(0)} = 5$ hence $R^{(0)} = 1$, see (2.15), (2.16), and $\lambda = 1/6$, $\mu = 1/3$, see (3.3 (b), (c))) reads

$$w_1 - i(\omega/6)w - i(\omega/3)xw_x = a_{0001} w_{xxx} + b_{201} w^2 w_{xx} + b_{12} w w_x^2 + c_{41}^{(0)} w^4 w_x. \tag{2.20}$$

This evolution PDE, (2.20), is C-integrable if $b_{12} = 3b_{201}$ and $c_{41}^{(0)} = -a_{0001}(b_{201})^2/3$. [6]. For arbitrary values of all the constants it features (with the only restriction that ω be real and nonvanishing, indeed, without loss of generality, *positive*, $\omega > 0$) this evolution PDE, (2.20), possesses many *completely periodic* solutions, with period $T = 12\pi/\omega$.

Another example of an evolution PDE belonging to the class (2.10) (with $p = 1, P = 3, q = 2, Q = 2, r^{(0)} = 3$ hence $R^{(0)} = 1, \lambda = \mu = 1/3$) reads

$$w_t - i(\omega/3)w - i(\omega/3)xw_x = a_{0001}w_{xxx} + b_{101}ww_{xx} + b_{02}w_x^2 - c_{21}w^2w_x.$$
$$(2.21)$$

For $b_{1001} = b_{02} = 0$ it is S-integrable (related by a change of variables to the modified Korteweg–de Vries equation). Again, it generally features many *completely periodic* solutions, with period $T = 6\pi/\omega$.

An example of an evolution PDE belonging to the class (3.2) (with $p = 2$) reads

$$w_{tt} - i(5/2)\Omega w_t - (3/2)\Omega^2 w$$
$$= a_2 w^2 + a_{11}ww_x + a_{02}w_x^2 + a_{101}ww_{xx} + a_{011}w_x w_{xx} + a_{002}w_{xx}^2.$$
$$(2.22)$$

Hence it features a lot of *completely periodic* solutions with period $T = 4\pi/\Omega$.

2.4 Proofs

To obtain (2.8), we start from the evolution PDE

$$\varphi_\tau = F(\varphi, \varphi_x, \varphi_{xx}, \dots)$$
$$(2.23)$$

with

$$F(\varphi, \varphi_x, \varphi_{xx}, \dots) = \sum_{p_0, p_1, p_2, \dots = 0, \sum_{n=0} p_n = p}^{p} a_{p_0 p_1 p_2 \dots}(\varphi)^{p_0}(\varphi_x)^{p_1}(\varphi_{xx})^{p_2} \dots$$
$$(2.24)$$

so that F satisfy the scaling property

$$F(\alpha\varphi, \alpha\varphi_x, \alpha\varphi_{xx}, \dots) = \alpha^p F(\varphi, \varphi_x, \varphi_{xx}, \dots).$$
$$(2.25)$$

In (2.23) $\varphi \equiv \varphi(x, \tau)$ is of course the dependent variable, while x and τ are the independent variables.

Now we use the change of dependent variables (2.1) with (2.2), (2.3) (2.4) (but with $\mu = 0$, see (2.5)), namely

$$w(x, t) = \exp(i\lambda\omega t)\varphi(x, \tau)$$
$$(2.26)$$

with

$$\tau[\exp(I\omega T) - 1]/(I\omega),$$
$$(2.27)$$

and we note that it entails (see (2.3))

$$w_t = i\lambda\omega w + \exp[i(\lambda + 1)\omega t]\varphi_\tau, \qquad (2.28)$$

hence, via (2.23), (2.25) and (2.26),

$$w_t - i\lambda\omega w = \exp\{i[\lambda(1 - p) + 1]\omega t\}F(w, W_x, w_{xx}, \dots). \qquad (2.29)$$

We now set

$$\lambda - 1/(p - 1), \qquad (2.30)$$

$$\Omega = \lambda\omega = \omega.(p - 1), \qquad (2.31)$$

and via (2.24) we get (2.8), and we thereby justify the assertions made above (after (2.9)) about the evolution PDE (2.8).

The proof of (2.9) is analogous: by differentiating (2.28) we get

$$w_{tt} = i\lambda\omega w_t + i(\lambda + 1)\omega \exp[i(\lambda + 1)\omega t]\varphi_\tau + \exp[i(\lambda + 2)\omega t]\varphi_{tt}, \quad (2.32)$$

and via (2.28) this yields

$$w_{tt} = i\lambda\omega w_t + i(\lambda + 1)\omega[w_t - i\lambda\omega w] + \exp[i(\lambda + 2)\omega t]\varphi_{tt}. \qquad (2.33)$$

We now assume that $\varphi \equiv \varphi(x, \tau)$ satisfy the *second-order* evolution PDE

$$\varphi_{tt} = F(\varphi, \varphi_x, \varphi_{tt}, \dots) \qquad (2.34)$$

rather than the *first-order* evolution PDE (4.1), with $F(\varphi, \varphi_x, \varphi_{xx}, \dots)$ defined as above, see (2.24), and therefore satisfying the scaling property (2.25). Hence we get

$$w_{tt} - i(2\lambda + 1)\omega w_t - \lambda(\lambda + 1)\omega^2 = \exp\{i[\lambda(1 - p) + 2]\omega t\}F(w, w_x, w_{xx}, \dots). \qquad (2.35)$$

We now set

$$\lambda 2/(p - 1), \qquad (2.36)$$

$$\Omega = \lambda\omega = 2\omega/(p - 1), \qquad (2.37)$$

and via (2.24) we get (2.9) and we thereby justify the assertions made above (after (2.9)) about the evolution PDE (2.9).

The derivations of (2.10) and (2.11) are analogous, except that we now use the change of variables (2.1) with nonvanishing μ, see (2.5). The relevant relations implied by this change of variables read then as follows:

$$w_t = i\lambda\omega w + i\mu\omega \exp(i\mu\omega t)x\varphi_\xi + \exp[i(\lambda + 1)\omega t]\varphi_t, \qquad (2.38)$$

as well as

$$w_x = \exp(i\mu\omega t)\varphi_\xi \tag{2.39}$$
$$w_{xx} = \exp(2i\mu\omega t)\varphi_{\xi\xi} \tag{2.40}$$

and so on, hence

$$w_t - i\lambda\omega w - i\mu\omega x w_x = \exp[i(\lambda+1)\omega t]\varphi_\tau \tag{2.41}$$

We now assume that $\varphi \equiv \varphi(\xi, \tau)$ satisfies the evolution PDE

$$\varphi_\tau = F^{(p)}(\varphi, \varphi_x, \varphi_{xx}, \dots) + F^{(q)}(\varphi, \varphi_x, \varphi_{xx}, \dots)$$
$$+ \sum_{j=0} F^{r^{(j)}}(\varphi, \varphi_x, \varphi_{xx}, \dots) \tag{2.42}$$

with

$$F^{(p)}(\varphi, \varphi_\xi, \varphi_{\xi\xi}), \dots)$$
$$= \sum_{\substack{p_0, p_1, p_2, \dots = 0 \\ \sum_n p_n = p, \ \sum_{n=1} n p_n = P}}^{p} a_{p_0 p_1 p_2 \dots}(\varphi)^{p_0}(\varphi_\xi)^{p_1}(\varphi_{\xi\xi})^{p_2 \dots} \tag{2.43}$$

and $F^{(q)}$ respectively $F^{(r^{(j)})}$ defined by analogous formulas except for the systematic replacement of the letters p, P and a by q, Q and b respectively by $r^{(j)}$, $R^{(j)}$ and $c^{(j)}$ (preserving of course the integer subscripts wherever they appear). Note that (2.44) and its analogs entail the scaling properties

$$F^{(p)}(\alpha\varphi, \alpha\beta\varphi_\xi, \alpha\beta^2\varphi_{\xi\xi}) = \alpha^p\beta^P F(\varphi, \varphi_\xi, \varphi_{\xi,\xi}, \dots), \tag{2.44}$$
$$F^{(q)}(\alpha\varphi, \alpha\beta\varphi_\xi, \alpha\beta^2\varphi_{\xi\xi}) = \alpha^q\beta^Q F(\varphi, \varphi_\xi, \varphi_{\xi,\xi}, \dots), \tag{2.45}$$
$$F^{(r^{(j)})}(\alpha\varphi, \alpha\beta\varphi_\xi, \alpha\beta^2\varphi_{\xi\xi}) = \alpha^{r^{(j)}}\beta^{R^{(j)}} F(\varphi, \varphi_\xi, \varphi_{\xi,\xi}, \dots). \tag{2.46}$$

Hence, from (2.41), (2.42), (2.1), (2.39), (2.40), (2.41), (2.44), (2.45), (2.46), (2.44) and the analogous equations to (2.44), we get (2.10), provided there hold the following relations:

$$\lambda(p-1) + \mu P = 1, \tag{2.47}$$
$$\lambda(q-1) + \mu Q = 1, \tag{2.48}$$
$$\lambda(r^{(j)}-1) + \mu R^{(j)} = 1. \tag{2.49}$$

It is then easily seen that, provided the condition (2.12) holds, the two equations (2.48), (2.49) yield (2.15), (2.16); and by then inserting these expressions, (2.15), (2.16), of λ and μ in (2.49) one obtains (2.14). The derivation of (2.10) is thereby completed, and via this derivation the statements made above (after (2.11)) about the evolution PDEs (2.10) are validated as well.

The proof of (2.11) is entirely analogous, except for the replacement of the *first-order* evolution PDE (2.34) that served as starting point for the derivation of (2.10), (2.11) with the *second-order* evolution PDE that obtains from (2.34) by replacing, in its left-hand side, φ, with $\varphi_{\tau\tau}$. The detailed treatment is left as an easy exercise for the diligent reader.

Let us end this section by displaying, for some of the above evolution PDEs, certain simple solutions which indeed feature the periodicity properties mentioned above (but are of course not the only ones to possess this property).

Consider the evolution PDE

$$w_t - i\Omega w = \sum_{\substack{p_0,p_1,p_2,\ldots=0 \\ \sum_{n=0}^{p} p_n = p, \; \sum_{n=1}^{p} np_n = p}}^{p} a_{p_0 p_1 p_2 \ldots} (w)^{p_0} (w_x)^{p_1} (w_{xx})^{p_2} \cdots \quad (2.50)$$

which is clearly a subcase of (2.8) (due to the *additional* restriction $\sum_{n=1}^{p} np_n = P$ on the summations indices p_n). Here p is an arbitrary *positive integer* larger than unity, $p > 1$, P is an arbitrary *nonnegative integer* different from unity, $P \neq 1$, the constants a_{p_0,p_1,p_2} are *arbitrary* (possible complex), and $\Omega > 0$.

It is then easily seen that this evolution PDE, (2.50), possesses the solution

$$w(x,t) = \exp(i\Omega t)\{Ax + B\exp[i(p-1)\Omega t] + C\}^P, \quad (2.51)$$

$$\beta = (P-1)/(p-1), \quad (2.52)$$

$$B = -i[(P-1)\Omega]^{-1}A^P \times$$
$$\sum_{\substack{p_0,p_1,p_2,\ldots=0 \\ \sum_{n=0}^{p} p_n = p \\ \sum_{n=1}^{p} np_n = P}}^{p} a_{p_0 p_1 p_2 \ldots} \prod_{n=1} [\beta(\beta-1)\cdots(\beta-n)]^p, \quad (2.53)$$

with A and C arbitrary (complex) constants. This solution is periodic in t with period $2\pi/\Omega$ provided $x \neq x_\pm$, while it becomes singular at the *real* times $t = t_\pm \mod \{2\pi/[(p-1)\Omega]\}$ for $x = x_\pm$, with

$$x_\pm = -\mathrm{Re}(AC^*) \pm \left\{ \left[\mathrm{Re}(AC^*)\right]^2 + |AB|^2 \right\}^{1/2} \tag{2.54}$$

$$\exp\left[i(p-1)\Omega t_\pm\right] = -(Ax_\pm + C)/B. \tag{2.55}$$

Here we are of course assuming that B, see (2.53), does not vanish, $B \neq 0$.

Likewise, consider the evolution PDE

$$w_{tt} - i\left[(p+3)/2\right]\Omega w_t - \left[(p+1)/2\right]\Omega^2 w$$

$$= \sum_{\substack{p_0,p_1,p_2,\ldots=0 \\ \sum_{n=0}^{p} p_n = p, \, \sum_{n=1} n p_n = P}}^{p} a_{p_0 p_1 p_2 \cdots}(w)^{p_0}(w_x)^{p_1}(w_{xx})^{p_2} \cdots \tag{2.56}$$

which is clearly a *subcase* of (2.9) (due to the *additional* restriction $\sum_{n=1} n p_n = P$ on the summations indices p_n). Here p is again an arbitrary *positive integer* different from unity, $p > 1$, P is an arbitrary *nonnegative integer* different from 2, $P \neq 2$, the constants $a_{p_0, p_1 p_2 \cdots}$ are *arbitrary* (possibly complex), and $\Omega > 0$.

It is then easily seen that this evolution PDE, (2.56), also possesses a solution analogous to (2.54), namely

$$w(x,t) = \exp(i\Omega t)\left\{ Ax + B\exp\left[i(p-1)\Omega t\right] + C\right\}^\beta, \tag{2.57}$$

$$\beta = (P-2)/(p-1), \tag{2.58}$$

$$B = \pm 2i\left[(P-2)(P-p-1)\Omega\right]^{-1}A^{p/2}$$

$$\times \left\{ \sum_{\substack{p_0,p_1,p_2,\ldots=0 \\ \sum_{n=0}^{p} p_n = p, \, \sum_{n=1} n p_n = P}}^{p} a_{p_0 p_1 p_2 \cdots} \prod_{n=1} [\beta(\beta-1)\cdots(\beta-n)]^p \right\}^{1/2},$$

$$\tag{2.59}$$

with A and C again arbitrary (complex) constants. This solution has of course the same periodicity properties described above, see after (2.54).

It is clearly easy to identify subclasses of (2.10) and (2.11) that possess solutions analogous to (2.54), (2.57), namely

$$w(x,t) = \exp(i\lambda w t\left\{ Ax\exp(i\mu\omega t) + B\exp(i\omega t) + C\right\}^\beta. \tag{2.60}$$

We leave this task, and the discussion of the properties of such solutions, as an exercise for the diligent reader.

References

[1] F. Calogero, A class of integrable hamiltonian systems whose solutions are (perhaps) all completely periodic, *J. Math. Phys.* **38** (1997), 5711-5719.

[2] F. Calogero and J.-P. Françoise, Solution of certain integrable dynamical systems of Ruijsenaars–Schneider type with completely periodic trajectories, *Ann. Henri Poincaré* **1** (2000), 173-191.

[3] F. Calogero, Classical many-body problems amenable to exact treatments, *Lecture Notes in Physics Monograph* **m 66**, Springer, 2001.

[4] F. Calogero and J.-P. Françoise, Periodic solutions of a many-rotator problem in the plane, *Inverse Problems*, (in press).

[5] F. Calogero and J.-P. Françoise, Periodic motions galore: how to modify nonlinear evolution equations so that they feature a lot of periodic solutions, (in preparation).

[6] F. Calogero, The evolution PDE ut = uxxx + 3(uxxu2+3ux2u) + 3uxu4, *J. Math. Phys.* **28** (1987), 538-555.

[7] F. Calogero, Why are certain nonlinear PDEs both widely applicable and integrable?, in: *What is integrability?* (V. E. Zakharov, editor), Springer, 1990, pp.1-62.

3
Geometry and integrability

Ron Y. Donagi

University of Pennsylvania

3.1 Introduction

These lectures are centered around the following result and its various special cases, applications, and extensions:

Theorem. *There is an algebraically integrable system on the moduli space of meromorphic Higgs bundles on a curve.*

This was proved independently by Markman [M] and Bottacin [Bo], and is closely related to results of Mukai [Mu] and Tyurin [T]. It incorporates and generalizes earlier work of Hitchin [H] and many others. The theorem combines ideas from algebraic geometry and symplectic geometry. In keeping with the expository aim of the lectures, the bulk of these notes concerns not the theorem and its applications, but the many ingredients which go into its proof. It is my hope that students with a fairly modest background in geometry will be able to work through these notes, learning a fair amount of algebraic geometry and symplectic geometry along the way. They may also be motivated to follow some of the leads in the last section towards open problems and further development of the subject.

The symplectic geometry needed for the statement and proof of the theorem is covered in Sections 3.2, 3.3, and 3.7, while the algebraic geometry is in Sections 3.4, 3.5, 3.6. Section 3.2 introduces the basics of symplectic and Poisson manifolds, while Section 3.3 discusses integrable systems. The notions of moment map and symplectic reduction, which are used in the proof, are explained in Section 3.7. The main algebraic geometry input is the study, in Section 3.5, of various aspects of the moduli of vector bundles and related objects such as principal G-bundles and Higgs bundles. Since vector bundles exhibit quite a range of complicated behavior, I preceded this section with a review of the much simpler story

21

for line bundles, in Section 3.4. Finally, Section 3.6 contains the specifics about spectral and cameral covers which are needed for exhibiting the Casimirs and Hamiltonians of the integrable system of the theorem.

In some ways, these notes are an elementary introduction to the more complete earlier version [DM]. But I have also taken this opportunity to update some of the results of [DM] and to point out their recent variations and applications. This is done mostly in Section 3.8. Many examples and special cases of the theorem are discussed there, together with various applications in mathematics and physics, further developments and some open problems. However, since these notes were getting to be too long, this discussion is not as leisurely as most of the rest of the text. Instead, only the main points of each special case, application, or open problem are explained, and the interested reader is referred to the literature for more details.

It is a pleasure to thank Yavuz Nutku and the Gursey Institute for the invitation to deliver these lectures and to write these notes. I would also like to acknowledge partial support from NSF Grant # DMS-9802456.

3.2 Symplectic geometry

3.2.1 Symplectic manifolds

A *Symplectic manifold* (M, ω) consists of:

M: a C^∞ manifold,

$\omega \in \mathcal{A}^2(M) := \Gamma(M, \Lambda^2 T^*(M)$: a closed, nondegenerate 2-form on M.

Note: any 2-form ω determines an interior (or: contraction) map $i_\omega : TM \to T^*M$. We say ω is nondegenerate if i_ω is an isomorphism. This is possible only when $dim M = 2n$ is even. Darboux's theorem guarantees the existence of local coordinates $p_i, q^i (i = 1, ..., n)$ on M such that locally

$$\omega = \Sigma_{i=1}^n dp_i \wedge dq^i.$$

A *Holomorphically symplectic manifold* (M, ω) consists of:

M: a complex (analytic) manifold

ω: a closed, non-degenerate holomorphic two form on M.

Note: saying that ω is holomorphic means that in terms of local holomorphic coordinates z^i it can be written as $\omega = \Sigma f_{ij}(z)dz^i \wedge dz^j$

(f_{ij} is holomorphic and there are no $d\bar{z}^i$.) The holomorphic version of Darboux's theorem says that for a holomorphically symplectic ω there are local holomorphic coordinates p_i, q^i such that $\omega = \Sigma_{i=1}^n dp_i \wedge dq^i$. The complex dimension of X must be even, the real dimension is divisible by 4.

3.2.2 Examples

There are two basic examples:

(i) For any C^∞ manifold X, the cotangent bundle $M = T^*X$ is symplectic. A choice of local coordinates q^i, $i = 1, \ldots, n$ on X determines pullback functions, still denoted q^i, on M, as well as fiber coordinates p_i along the cotangent spaces. The locally defined two form $\omega = \Sigma dp_i \wedge dq^i$ is easily seen to be independent of the choice of local coordinates, so it is a (global) symplectic form on M. In fact the 1-form $\alpha := \Sigma p_i dq^i$ is already globally defined, so $\omega = d\alpha$ is actually exact.

If we start instead with a complex manifold X, the same construction produces a holomorphically symplectic manifold (M, ω) where $M := T^*X$ is the cotangent bundle and ω is as above.

(ii) Let G be a Lie group and $\mathbf{g}$ its Lie algebra. The action of G on itself, by conjugation, induces the adjoint action of G on $\mathbf{g}$ and the coadjoint action on the dual vector space $\mathbf{g}^*$:

$$Ad : G \times \mathbf{g} \to \mathbf{g}, \qquad Ad^* : G \times \mathbf{g}^* \to \mathbf{g}^*.$$

For each $\xi \in \mathbf{g}^*$ consider its orbit $\mathcal{O} = \mathcal{O}_\xi := G \cdot \xi \subset \mathbf{g}^*$ under the coadjoint action. There is a natural symplectic form ω on $\mathcal{O}$ (discovered by Kirillov and Kostant), making $\mathcal{O}$ into a symplectic manifold for real G and a holomorphically symplectic manifold for complex G.

Explicitly, we use the map $G \to \mathcal{O}$ sending $g \mapsto g\xi$ in order to identify the tangent space $T_\xi \mathcal{O}$ to $\mathcal{O}$ at each $\xi \in \mathcal{O}$ with the quotient $\mathbf{g}/\mathbf{g}_\xi$, where $G_\xi := \{g \in G|\ g\xi = \xi\}$ is the stabilizer of ξ in G, and $\mathbf{g}_\xi = \{X \in \mathbf{g}|\ (\xi, [X, Y]) = 0\ , \forall Y \in \mathbf{g}\}$ is the Lie algebra of G_ξ, a subalgebra of $\mathbf{g}$. With this identification, let $\bar{X}, \bar{Y} \in T_\xi \mathcal{O} \approx \mathbf{g}/G^*$ be the images of $X, Y \in \mathbf{g}$. The symplectic form is then defined by

$$\omega : \bar{X}, \bar{Y} \mapsto (\xi, [X, Y]),$$

which of course is independent of the representatives X, Y used.

Note: when G is semisimple, the Killing form gives an isomorphism of $\mathbf{g} \widetilde{\rightarrow} \mathbf{g}^*$ which is G-equivariant. So coadjoint orbits can be naturally identified with adjoint orbits, in $\mathbf{g}$.

Example. The Lie algebra of $G = SO(3, R)$ is

$$\mathbf{g} = so(3, R) \approx R^3 \approx \left\{ \begin{pmatrix} 0 & a & c \\ -a & 0 & b \\ -c & -b & 0 \end{pmatrix} \,\middle|\, a, b, c \in R \right\}.$$

Here co-adjoint orbit = adjoint orbit = sphere $\{a^2 + b^2 + c^2 = r^2\}$ in R^3.

3.2.3 Poisson manifolds

A Poisson manifold (M, ψ) consists of a manifold M with a 2-vector $\psi \in \Gamma(\Lambda^2 TM)$ such that the (Poisson) bracket

$$f, g \in C^\infty(M) \mapsto \{f, g\} := (df \wedge dg, \psi)$$

is a Lie algebra bracket on $C^\infty(M)$, i.e it satisfies the Jacobi identity

$$\{f, \{g, h\}\} + \{g, \{h, f\}\} + \{h, \{f, g\}\} = 0.$$

In this case the map

$$v : C^\infty(M) \to VF(M) := \Gamma(M, TM),$$

$$f \mapsto (df, \psi) = i_{df} \psi$$

is automatically a homomorphism of Lie algebras:

$$v(\{f, g\}) = [v(f), v(g)].$$

The vector field $v(f)$ is called the Hamiltonian vector field of the function f, while f is referred to as the Hamiltonian function of $v = v(f)$.

Examples

- A symplectic manifold (M, ω) is Poisson: the 2-vector ψ is determined by the isomorphism $i_\psi : T^*M \to TM$ which is the inverse of $i_\omega : TM \widetilde{\rightarrow} T^*M$. The Jacobi identity for ψ turns out to be equivalent to the closedness of ω. So a Poisson manifold is symplectic if and only if ψ is non-degenerate.

- The dual $\mathbf{g}^*$ of a Lie algebra $\mathbf{g}$ is a Poisson manifold. For $F, G \in C^\infty(\mathbf{g}^*)$ and $\xi \in \mathcal{G}^*$, the bracket is defined by

$$\{F, G\}_\xi := \left(\xi, [d_\xi F, d_\xi G]\right).$$

Note that in this case the 2-vector ψ is degenerate. Its restriction to each coadjoint orbit $\mathcal{O} \subset \mathbf{g}^*$ is non-degenerate and corresponds precisely to the standard symplectic form on $\mathcal{O}$ described earlier. In fact, the coadjoint orbits are the *largest* loci on which ψ is non-degenerate: the conormal space to the orbit $\mathcal{O}_\xi$ at ξ is precisely the nullspace of ψ at ξ.

3.2.4 Symplectic leaves

A general result of Weinstein asserts that any Poisson manifold (M, ψ) is the disjoint union of the submanifolds of M on which ψ is non-degenerate and which are maximal with respect to this property. These submanifolds inherit a symplectic structure and they are called the *symplectic leaves* of (M, ψ). Their conormal space at each point is the nullspace of ψ there. The symplectic leaves of $\mathbf{g}^*$ with the Kirillov–Kostant Poisson structure are, of course, the coadjoint orbits. For $so(3, R)$ the picture is very simple: the symplectic leaves are the spheres centered at the origin, as well as the origin itself. Another extreme is the case where ψ is algebraic and non-degenerate somewhere: there is then an open symplectic "leaf", and possibly others, of lower dimension, in its closure. In general, the rank of the alternating 2-vector ψ is not constant but only semicontinuous, i.e. its value at a point is less than or equal to its values at nearby points; this rank equals the dimension of the symplectic leaves, which can be lower for a special leaf than for nearby ones.

3.3 Integrable systems

3.3.1 Definitions

The notion of a *Hamiltonian map* gives a coordinate-free way to discuss a collection of commuting Hamiltonians. A map $H : (M, \psi) \to B$ from a Poisson manifold (M, ψ) to another manifold B is called Hamiltonian if for every two functions $f, g \in C^\infty(B)$, the pullback functions $H^* f := f \circ H$ and $H^* g := g \circ H$ Poisson commute:

$$\{H^* f, H^* g\} = 0.$$

A *Casimir* on the Poisson manifold (M, ψ) is a function $f \in C^\infty M$ whose Hamiltonian vector field vanishes: $v(f) = 0$. Equivalently, f Poisson-commutes with every function on M. More generally, a map $M \to C$ is called Casimir if the pullback of any function on C is a Casimir on (M, ψ).

An *integrable system* is a Hamiltonian map of maximal rank. When M is symplectic of dimension $2n$, this amounts to a Hamiltonian map $H : M \to B$ which is onto an n-dimensional base B. More generally, let (M, ψ) be Poisson with ψ of constant rank $2n$, where $dim(M) = 2n + c$. Then we want $B = H(M)$ to have dimension $c + n$. Locally, then, H can be expressed in terms of $n + c$ independent functions on M. Of these, c will be Casimirs and the remaining n Hamiltonians need to Poisson-commute.

3.3.2 Liouville's Theorem

Let $H : (M, \psi) \to B$ be an integrable system where

$$dim(M) = 2n + c,$$

$$rank(\psi) = 2n,$$

$$dimB = n + c,$$

H is a proper, submersive, Hamiltonian map.

Then:

- The connected components of the fibers of H are tori, i.e. they are diffeomorphic to $(S^1)^n = R^n / \mathbf{Z}^n$.
- The Hamiltonian vector fields $v(H^* f)$ for $f \in C^\infty(B)$ are translation invariant vector fields on these tori, so the corresponding flows are linear.
- The symplectic foliation of M is (locally) pulled back from a foliation of B.

Note: saying that H is submersive means that its differential

$$dH : T_m M \to T_b B, \quad b := H(m)$$

is surjective for all $m \in M$. This guarantees that the fibers $H^{-1}(b)$ are non-singular manifolds. In practice, an integrable system is often submersive only over some dense open subset of the base B, and Liouville's theorem applies to the fibers over this subset.

3.3.3 Algebraically integrable systems

The most natural way to complexify the notion of an integrable system is to turn Liouville's theorem into a definition. Thus an *analytically integrable system* consists of a complex analytic manifold M, an analytic Poisson structure ψ, (i.e. a holomorphic two-vector on M satisfying the Jacobi identity), a proper, holomorphic Hamiltonian map $H : (M, \psi) \to B$ whose generic fibers are complex tori $\mathbf{C}^n / \Lambda$, where $\Lambda \approx \mathbf{Z}^{2n}$ is a maximal lattice in $\mathbf{C}^n$. An *algebraically integrable system* involves data $(M, \psi, H, B, ...)$ which are algebraic. In particular, the complex tori now become *abelian* varieties. This restricts the lattices Λ which may arise: they must satisfy Riemann's first and second bilinear relations, cf. [GH]. Note that we do not require H to be submersive. Accordingly, we expect the generic fiber (i.e. the fiber over b in a Zariski open subset of B) to be a complex torus or abelian variety, but we allow some of the fibers to degenerate. Once a system is shown to be algebraic, it can be 'solved' explicitly: the solutions are flows which are tangent to an abelian variety fiber. On the universal cover $\mathbf{C}^n$ this flow is linear; on the abelian variety itself it can therefore be expressed in terms of theta functions.

3.4 Line bundles

Before discussing moduli spaces of vector bundles and more complicated objects, it may make sense to review the much simpler case of line bundles. In this section we work over a fixed (compact, non-singular) Riemann surface X. Since we are switching here from symplectic to algebraic language, we think of X as a 1-dimensional algebraic variety over $\mathbf{C}$, or a "curve" for short. In fact much of what we say will work over a (projective, non-singular) complex variety X of any dimension.

A vector bundle on X is another variety V, together with a map $\pi : V \to X$ which is locally isomorphic to the product $X \times \mathbf{C}^n$ (with the projection map): every point $x \in X$ has a neighborhood $\mathcal{U}$ and a trivialization $\pi^{-1}(\mathcal{U}) \widetilde{\to} \mathcal{U} \times \mathbf{C}^n$ which commutes with projection to $\mathcal{U}$ (i.e. it sends fibers to fibers), and on the intersection $\mathcal{U}_1 \cap \mathcal{U}_2$ of two such neighborhoods, the difference between these trivializations is linear in the fibers and algebraic (or holomorphic) along the base, i.e it is given by an $n \times n$ transition matrix $g(x)$ whose entries are algebraic (or holomorphic) functions of $x \in \mathcal{U}_1 \cap \mathcal{U}_2$. The vector bundles we defined here are algebraic (or holomorphic; it turns out not to matter). There is an analogous, and more familiar, notion of a C^∞ complex vector bundle where the matrices $g(x)$ are allowed to be complex-valued C^∞ functions of x.

3.4.1 Pic and Jac

A line bundle is a vector bundle of rank $n = 1$. The moduli space of line bundles is usually called the Picard variety, and denoted $Pic(X)$. It is an algebraic variety whose points are in one to one correspondence with the isomorphism classes of line bundles on X. It is also a group, the operation being the tensor product of line bundles. As a variety, $Pic(X)$ is disconnected. The connected component of the trivial bundle is called the Jacobian, and is denoted by $Jac(X)$. The map which sends a line bundle to its first Chern class, or degree, is continuous (i.e. constant on connected components) and is a surjecive group homomorphism $Pic(X) \to \mathbf{Z}$. Its kernel is $Jac(X)$. Algebraic line bundles have more structure than C^∞ bundles: two algebraic line bundles are isomorphic as C^∞ bundles if and only if their Chern classes agree. The Jacobian therefore parametrizes all possible algebraic (or holomorphic) structures on the trivial C^∞ bundle. An analogous description still holds for the Picard variety parametrizing line bundles on higher dimensional (non-singular, projective) X, the main difference being that the group of connected components of $Pic(X)$ (or the group of isomorphism classes of the underlying C^∞ line bundles) can be bigger that $\mathbf{Z}$.

3.4.2 Cohomological description

The description of vector bundles in terms of their transition matrices can be refined to show that an isomorphism class of vector bundles is uniquely determined by a cohomology class in $H^1(X, GL_n(\mathcal{O}_X))$. For line bundles, the sheaf $GL_1(\mathcal{O}_X) = \mathcal{O}_X^*$ is just the sheaf of nowhere vanishing holomorphic (or algebraic) functions on X, so we can identify $Pic(X)$ with $H^1(X, \mathcal{O}_X^*)$. This can be described via the *exponential sequence* of sheaves on X:

$$0 \to \mathbf{Z} \to \mathcal{O}_X \xrightarrow{\exp} \mathcal{O}_X^* \to 1.$$

Indeed, the corresponding long exact sequence of cohomology gives

$$0 \to H^1(X, \mathcal{O}_X)/H^1(X, \mathbf{Z}) \to H^1(X, \mathcal{O}_X^*) \xrightarrow{c_1} H^2(X, \mathbf{Z}) \to 0$$

which can be identified with the previous sequence

$$0 \to Jac(X) \to Pic(X) \xrightarrow{c_1} \mathbf{Z} \to 0.$$

(For higher dimensional X, the group $H^2(X, \mathbf{Z})$ can be bigger than $\mathbf{Z}$ and the image of the Chern class map c_1 may be a proper subgroup: it is the kernel of $H^2(X, \mathbf{Z}) \to H^2(X, \mathcal{O}_X)$.)

Let X be a (smooth, compact) curve of genus g. Then $H^1(X, \mathbf{Z})$ is a free abelian group of rank $2g$ and $H^1(X, \mathbf{C})$ is its complexification, a $2g$-dimensional complex vector space. The Hodge theorem says that $H^1(X, \mathbf{C})$ decomposes as the sum of two g-dimensional complex subspaces, $H^{1,0} \oplus H^{0,1}$, where

$$H^{1,0} = H^0(X, K_X)$$

$$H^{0,1} = H^1(X, \mathcal{O}_X),$$

namely, the spaces of holomorphic and anti-holomorphic 1-forms on X. (We use K_X as another notation for the *canonical bundle* of X, $K_X = T_X^*$, and we denote the *trivial bundle* by $\mathcal{O}_X$.) The map $H^1(X, \mathbf{Z}) \to H^1(X, \mathcal{O}_X)$ is the composition of the inclusion $H^1(X, \mathbf{Z}) \hookrightarrow H^1(X, \mathbf{C})$ with the projection $H^1(X, \mathbf{C}) \to H^{0,1} = H^1(X, \mathcal{O}_X)$. Since complex conjugation is an automorphism of $H^1(X, \mathbf{C})$ which fixes $H^1(X, \mathbf{Z})$ but interchanges $H^{1,0}$ with its orthogonal complement $H^{0,1}$, it follows that the image of $H^1(X, \mathbf{Z})$ in $H^1(X, \mathbf{C})$ does not intersect $H^{1,0}$. The composition $H^1(X, \mathbf{Z}) \to H^1(X, \mathcal{O}_X)$ is therefore injective, i.e. $H^1(X, \mathbf{Z})$ is a maximal lattice in the g-dimensional vector space $H^1(X, \mathcal{O}_X)$. We conclude that $Jac(X) = H^1(X, \mathcal{O}_X)/H^1(X, \mathbf{Z})$ is a g-dimensional complex torus. (In fact, it is an abelian variety.) The same is true for the ith component $Pic^i(X)$ of $Pic(X)$, i.e. for the component mapping by c_1 to $i \in \mathbf{Z}$: it is (non-canonically) isomorphic to $Pic^0(X) = Jac(X)$. Such an isomorphism is given by tensoring with any fixed line bundle of degree i.

3.4.3 Flat bundles

A flat bundle is given by transition matrices $g(x)$ which are constant, i.e. independent of x. Such a bundle admits a natural flat connection: the flat sections are those which are locally constant in terms of any of the local trivializations, and this condition does not depend on the trivialization used since any two trivializations differ by a constant matrix. Conversely, specifying a bundle with a flat connection uniquely determines a flat bundle. Cohomologically, the moduli space of flat bundles on X is given by $H^1(X, \mathbf{C}^*)$. It is the product of $2g$ copies of $\mathbf{C}^*$, when X is a curve of genus g. Note that sections of the sheaf $\mathbf{C}^*$ are locally constant, non-zero functions, compared to sections of $\mathcal{O}_X^*$ which are holomorphic or algebraic non-zero functions. The inclusion $i : \mathbf{C}^* \hookrightarrow \mathcal{O}_X^*$ induces the homomorphism $i_* : H^1(X, \mathbf{C}^*) \to H^1(X, \mathcal{O}_X^*) - Pic(X)$ which sends a flat bundle to the bundle with the same algebraic transition matrices, forgetting that they actually happen to be constant.

Topologically, this map can be described as follows. Its image is $Pic^0(X) = Jac(X) \approx (S^1)^{2g}$, and the map $(\mathbf{C}^*)^{2g} \approx H^1(X, \mathbf{C}^*) \to Jac(X) \approx (S^1)^{2g}$ is the product of $2g$ copies of the argument map

$$\mathbf{C}^* \to S^1$$

$$z \mapsto z/|z|.$$

Thus, a line bundle $L \in Pic(X)$ admits a flat structure if and only if its degree is 0, and in this case the family of flat structures on L, or more simply the family of flat connections on L, is an affine space modelled on $H^0(X, K_X) = H^{1,0} \approx \mathbf{C}^g \approx R^{2g}$.

The general flat bundle has structure group (or holonomy group) $\mathbf{C}^*$. There is the special class of *unitary flat bundles*, for which the locally constant transition matrices take values in the unitary group $U(1) \approx S^1$. The moduli space of these unitary flat bundles is given by

$$H^1(X, U(1)) \approx (U(1))^{2g} \approx (S^1)^{2g}.$$

It is thus diffeomorphic to the Jacobian. The theorem of Narasimhan-Seshadri says (in this case) that any degree 0 holomorphic line bundle on X admits a unique flat unitary connection, i.e. the restriction of i_* to the moduli space of unitary flat bundles is an isomorphism to the Jacobian.

3.4.4 Abel–Jacobi

Any point p of the curve X determines a line bundle L_p as follows. We cover X by two open sets: a small disc $\mathcal{U}_p$ containing p, and the complement $X - p$. Let $z = z(x)$ be a local coordinate for $x \in \mathcal{U}_p$, vanishing at p. The line bundle L_p is trivialized on each of the open sets; on their intersection the 1×1 transition "matrix" g is taken to be the coordinate z. It is straightforward to check that this L_p depends only on $p \in X$ and not on the coordinate z. The fact that z vanishes at a single point implies that $degree(L_p) = 1$. We get this way a map

$$AJ : X \to Pic^1(X)$$

$$p \mapsto L_p$$

called the Abel-Jacobi map. It turns out to be an algebraic map, and much can be learned about the geometry of a curve from the behavior of its Abel-Jacobi map. But we will not pursue this here except for the following remarks:

- The choice of a *base point* $p_0 \in X$ allows us to view AJ as going to $Jac(X) = Pic^0(X)$ instead of $Pic^1(X)$: simply replace L_p by $L_p \otimes (L_{p_0})^{-1}$, where $(L_{p_0})^{-1}$ is the *inverse* or *dual* line bundle of L_{p_0}, obtained using the same open cover and the inverse transition function $g^{-1}(x)$.

- When $g = 0$ we have $Pic(X) \approx \mathbf{Z}$, $Jac(X)$ is a point and the Abel-Jacobi map is a constant. When $g = 1$, the Abel-Jacobi map is an isomorphism of X with $Jac(X)$ (or, more naturally, with $Pic^1(X)$). For $g \geq 2$ it is injective, but of course not surjective: Only some special line bundles can be described algebraically via a cover involving only two open sets as above.

3.5 Vector bundles

3.5.1 *Complications*

The set of isomorphism classes of rank n vector bundles on a curve X is given by the cohomology group $H^1(X, GL_n(\mathcal{O}_X))$. Algebraically this is more complicated than the case of line bundles because $GL_n(\mathcal{O}_X)$ is non-abelian for $n > 1$. But the main cause of complications in the theory is geometric: it is the existence of the *jump phenomenon*. There exist vector bundles V on the product $X \times \Delta$ of X with a parameter space Δ (which can be taken to be "a small disc") and two points $0, 1 \in \Delta$ such that the restriction V_δ of V to $X \times \delta$, for $\delta \in \Delta$, is isomorphic to the vector bundle V_1 for $\delta \neq 0$, but V_0 is not isomorphic to V_1. If V_0 and V_1 represent points of a moduli space of vector bundles, this example presents us with an (algebraic) map v from the parameter space Δ to the moduli space, which jumps: $v(\Delta - 0)$ is one point, while $v(0)$ is another point. In other words, the topology of the moduli space would be non-separated. (We will see an explicit example of such a jump at the end of this section.)

This forces us to accept some compromise. We could disallow some bundles and thus settle for a moduli space which parametrizes only some subset of all bundles, and thereby avoids the jump phenomenon. Or, we could allow certain non-isomorphic vector bundles to be represented by the same point of the moduli space: instead of excluding either V_0 or V_1, we allow both but declare them equivalent. A third possibility is to accept non-separatedness of the moduli space and to develop a language for studying it. All three approaches can be carried out: they lead to the moduli space of stable bundles, the moduli space of S-equivalence classes of semi-stable bundles, and the moduli stack of all bundles. We will describe some of their features below.

3.5.2 Stability

Just as for line bundles, a vector bundle has a well-defined degree or first Chern class. (One way to define this is to set $c_1(V) := c_1(\det V)$ where $\det V$ is the line bundle on X whose transition functions are the determinants of the transition matrices of V.) If V has rank n and degree d, we define its *slope* to be d/n: if V happens to be the direct sum $\oplus L_i$ of line bundles, then μ is the average degree of the L_i.

A bundle V is called *stable* if for every subbundle $V' \subset V$ (other than V itself and the zero subbundle),

$$\mu(V') < \mu(V).$$

Similarly V is *semistable* if for every subbundle $V' \subset V$,

$$\mu(V') \leq \mu(V).$$

S-equivalence is the equivalence relation on the set of isomorphism classes of semistable bundles generated by setting V_0 equivalent to V_1 whenever there is a jump from V_1 to V_0, as above. More concretely, consider a short exact sequence

$$0 \to V' \to V_1 \to V'' \to 0$$

where V', V'' are semistable bundles of the same slope. With V', V'' fixed, the extension is specified by an *extension class* $\epsilon \in H^1(X, (V'')^{-1} \otimes V')$. But since V', V'' themselves have a $\mathbf{C}^*$ of scalar automorphisms, it follows that the isomorphism class of the bundle V_1 depends on ϵ only up to multiplication by a non-zero scalar. By rescaling the extension class, we therefore obtain a jump from V_1, for all non-zero values of the scalar, to $V_0 := V' \oplus V''$ when the scalar becomes 0. In fact, it turns out that any two S-equivalent bundles can be linked by a sequence of moves of this particular type.

With these definitions, we can describe two of the three types of moduli spaces mentioned above. There exists a projective variety $\mathcal{M} = \mathcal{M}_X(n, d)$ parametrizing S-equivalence classes of semistable bundles of rank n and degree d on X. There is also a quasi-projective variety $\mathcal{M}^s = \mathcal{M}^s_X(n, d)$, identifiable with an open subset of $\mathcal{M}$, and parametrizing isomorphism classes of stable bundles on X. The latter is "almost" smooth: its singularities are quotient singularities, corresponding to bundles which have non-scalar automorphisms.

3.5.3 Tangent spaces

We can describe the tangent space to $\mathcal{M}^s$ at a non-singular point $V \in \mathcal{M}^s$ as follows. The bundle V is given in terms of some open cover $\{\mathcal{U}_i\}$ of X by a 1-cocycle $\{g_{ij}\}$, where g_{ij} is an $n \times n$ matrix of holomorphic functions on $\mathcal{U}_i \cap \mathcal{U}_j$. A tangent vector to $\mathcal{M}^s$ at V is determined by mapping a small disc $\Delta \subset \mathbf{C}$ to $\mathcal{M}^s$ so that $0 \in \Delta$ goes to V. This is achieved by deforming the transition matrices:

$$g_{ij}^{\epsilon}(x) = g_{ij}(x) + \epsilon g'_{ij}(x) + \cdots \qquad , \epsilon \in \Delta.$$

The tangent vector itself is then encoded in the 1-cocycle $\{g'_{ij}(x)\}$ giving the leading term of the deformation. The multiplicative cocycle condition for the bundle V:

$$g_{ij} \cdot g_{jk} = g_{ik}$$

then translates into an additive cocycle condition for $h_{ij} := \{g_{ij}^{-1} \cdot g'_{ij}\}$:

$$g_{jk}^{-1} \cdot h_{ij} \cdot g_{jk} + h_{jk} = h_{ik}.$$

The conjugation by g_{jk} means that the class of the deformation does not live in $H^1(X, gl_n(\mathcal{O}))$; rather, it lives in $H^1(X, adV)$. Here

$$adV = V^* \otimes V = gl(V)$$

is the bundle of endomorphisms of V. Its fiber at each point $x \in X$ is isomorphic to the Lie algebra gl_n, but the isomorphism is not natural and cannot be chosen globally over X. Anyway, the conclusion is that there is a natural identification

$$T_V \mathcal{M}^s \approx H^1(X, ad(V)),$$

or by Serre duality and the self-duality of $ad(V)$ (via the Killing form):

$$T_V^* \mathcal{M}^s \approx H^0(X, ad(V) \otimes K_X).$$

Note that in the abelian case $n = 1$, and only then, $ad(V)$ is trivial:

$$ad(V) - V \otimes V^* \approx \mathcal{O}_X \qquad \text{for } n = 1.$$

The result for vector bundles therefore specializes to:

$$T_L Pic(X) \approx H^1(X, \mathcal{O}_X),$$

which also follows trivially from the exponential sequence of Section 3.4.2.

The Riemann–Roch formula for a vector bundle W on a curve X of genus g says that:

$$\chi(W) := h^0(X, W) - h^1(X, W) = degree(W) - (g - 1) \cdot rank(W).$$

We apply this to $W := ad(V)$. This has rank n^2 and degree 0, so we find

$$\chi(ad(V)) = n^2(1 - g).$$

Now the only endomorphisms of a stable bundle V are scalar, thus we have $h^0(ad(V)) = 1$. This determines the dimension of $T_V \mathcal{M}^s \approx H^1(X, ad(V))$ and hence also of $\mathcal{M}^s$:

Corollary

$$dim \mathcal{M}_X^s(n, d) = (g - 1)n^2 + 1.$$

3.5.4 The moduli stack

We have considered the moduli space of stable bundles, in which some bundles have to be excluded for not being stable, as well as the moduli space of semistable bundles, in which fewer bundles are excluded, but some of these have to be identified with each other. The third approach is to insist on including all bundles. The resulting structure is no longer that of an algebraic variety or even a scheme, but an *algebraic stack* $M = M_X(n, d)$. We cannot discuss these here, so we only point out one of their features: each point $[V]$ of the stack (representing the isomorphism class of some vector bundle V) can be assigned a *dimension*, in such a way that whenever V_1 jumps to V_0 we have

$$dim[V_1] > dim[V_0].$$

In our case this assignment is straightforward: $dim[V]$ is defined to be

$$dim[V] := -dim\ (End(V)),$$

where $End(V)$ is the vector space of global endomorphisms of V, i.e. global sections of $ad(V)$. In general, the main difference between a stack and a scheme is that points of the stack "remember" the presence of some automorphisms. Locally, an algebraic stack can be described by an "equivalence relation" which may involve a continuous family of identifications of a family with itself.

The stack itself can be assigned a dimension. If it contains a Zariski open subset $\mathcal{U}$ which looks like an N-dimensional variety except that all its points have stack dimension $= -r$, then the dimension of the stack is $N - r$. For example, the moduli stack of vector bundles on X has dimension $(g - 1)n^2$, since stable bundles have a 1-dimensional family of automorphisms. This notion behaves well in fibrations: given a morphism of stacks $\pi : \mathcal{M} \to M$ whose fibers have dimension N, we have

$$dim(\mathcal{M}) = N + dim(M).$$

This is useful when M is the moduli stack of some objects V (e.g. bundles) while $\mathcal{M}$ is the moduli *space* of some enriched objects (V, δ) consisting of an object V plus some additional structure δ on V which is not preserved by any automorphism of V, so the pair (V, δ) has no automorphisms. Then $dim(\mathcal{M})$, as a variety, equals $dim(M)$, as a stack, plus the number of parameters required to specify δ.

3.5.5 Examples

First we consider vector bundles on $X = \mathbf{P}^1$. It is well known that every vector bundle on $\mathbf{P}^1$ is a direct sum of line bundles. Further, we have seen that $Jac(\mathbf{P}^1) = (0)$ and $Pic(\mathbf{P}^1) = \mathbf{Z}$. The line bundle of degree d on $\mathbf{P}^1$ is denoted $\mathcal{O}_{\mathbf{P}^1}(d)$, or $\mathcal{O}(d)$ for short. We can therefore enumerate all vector bundles on $\mathbf{P}^1$. For example, when $n = 2$ and the degree is $d = 0$, the possibilities are:

$$\mathcal{O} \oplus \mathcal{O}, \quad \mathcal{O}(1) \oplus \mathcal{O}(-1), \quad \mathcal{O}(2) \oplus \mathcal{O}(-2), \quad \ldots$$

We see immediately that none of these is stable, $\mathcal{O} \oplus \mathcal{O}$ is semistable, and the others are unstable because they contain a line bundle of positive slope (=degree). Therefore:

$$\mathcal{M}^s_{\mathbf{P}^1}(2,0) = \emptyset, \qquad \mathcal{M}_{\mathbf{P}^1}(2,0) = (point).$$

The moduli stack $M_{\mathbf{P}^1}(2,0)$ contains one point x_k for each $k \geq 0$, represented by the bundle $\mathcal{O}(k) \oplus \mathcal{O}(-k)$. An endomorphism of this bundle consists of a matrix $\begin{pmatrix} a & b \\ c & d \end{pmatrix}$ where a, d are sections of $\mathcal{O}_{\mathbf{P}^1}$ (i.e. complex numbers), b is a section of $\mathcal{O}_{\mathbf{P}^1}(2k)$ and c is a section of $\mathcal{O}_{\mathbf{P}^1}(-2k)$, hence $c \equiv 0$ unless $k = 0$. We find that:

$$dim(x_k) = \begin{cases} -4 & k = 0 \\ -(2k+3) & k \geq 1. \end{cases}$$

So the semistable point $\mathcal{O} \oplus \mathcal{O}$ has the largest dimension (namely, -4), and all others are smaller. In fact, the other points are in its closure. A similar picture holds for any even degree d. For odd d, $\mathcal{M}_{\mathbf{P}^1}(2,d)$ is empty while the stack $M_{\mathbf{P}^1}(2,d)$ still involves an infinite nested sequence of points $\mathcal{O}(k) \oplus \mathcal{O}(d-k)$.

Let V be a vector bundle of rank n and degree d on a curve X which is a branched cover $\pi : X \to Y$ of another curve Y, and let l be the degree of the cover π. There is a natural way to construct a *direct image* bundle $\pi_* V$ on Y. This is defined in such a way that for any open set $\mathcal{U} \subset Y$, sections of $\pi_* V$ on $\mathcal{U}$ correspond precisely to sections of V on $\pi^{-1}(\mathcal{U}) \subset X$. Thus if $y \in Y$ is a regular value (= not a branch point) of π and $\pi^{-1}(y) = \{x_1, ..., x_l\}$, then

$$(\pi_* V)_y = \oplus^l_{i=1} V_{x_i}.$$

There are three relations among the invariants of V and $\pi_* V$:

$$rank(\pi_* V) = l \cdot rank(V) = l \cdot n$$

$$h^0(\pi_* V) = h^0(V)$$

$$\bar{d} - l \cdot n \cdot (\bar{g} - 1) = d - n \cdot (g - 1).$$

In the last formula, g and $\bar{g}$ are the genera of Y and X respectively, and $d, \bar{d}$ are the degrees of $V, \pi_* V$. (This formula can best be remembered as stating that the holomorphic Euler characteristic is preserved under direct image.)

For example, if $X \approx Y \approx \mathbf{P}^1$ and $\pi : \mathbf{P}^1 \to \mathbf{P}^1$ is the standard double cover $z \mapsto w := z^2$, branched at 0 and ∞, the above formulas imply that $\pi_*\mathcal{O}(1)$ is a vector bundle on $\mathbf{P}^1$ of rank 2 and degree 0 with a 2-dimensional space of sections. In fact $\pi_*\mathcal{O}$ is $\mathcal{O} \oplus \mathcal{O}$ and a section of $\mathcal{O}(1)$ upstairs corresponds downstairs to the pair of sections representing its even and odd parts:

$$f(z) = f_+(w) + z\,f_-(w) \mapsto (f_+(w), f_-(w))$$

where

$$f_+(z^2) := \frac{1}{2}(f(z) + f(-z))$$

$$f_-(z^2) := \frac{1}{2z}(f(z) - f(-z)).$$

A more interesting example arises when Y is $\mathbf{P}^1$ and X is the elliptic curve

$$X : \{y^2 = x(x-1)(x-\lambda)\}$$

for some $\lambda \in \mathbf{C} - \{0, 1\}$, and $\pi : X \to Y$ sends $(x, y) \mapsto x$. For $L \in Pic^0(X)$ we find that π_*L is a rank 2 vector bundle of degree -2 on $\mathbf{P}^1$, and $h^0(\pi_*L) = h^0(L)$. For all L other than the trivial bundle $\mathcal{O}_X$, we get $h^0(\pi_*L) = 0$, so the only possibility is $\pi_*L \approx \mathcal{O}(-1) \oplus \mathcal{O}(-1)$. However, for $L = \mathcal{O}_X$, π_*L must have a non-zero section, and we see easily that in fact $\pi_*\mathcal{O}_X \approx \mathcal{O} \oplus \mathcal{O}(-2)$. This gives us an explicit example of the jump phenomenon: as L varies continuously in $Pic^2(L)$, its direct image π_*L varies continuously in the stack $M_{\mathbf{P}^1}(2, 0)$. It equals the semistable (or: generic) point $\mathcal{O}(-1) \oplus \mathcal{O}(-1)$ for most L, but jumps to the "smaller" point $\mathcal{O} \oplus \mathcal{O}(-2)$ at $L = \mathcal{O}_X$.

In order to see an example of S-equivalence, consider rank 2 bundles on an elliptic curve X. Any pair $L_1, L_2 \in Jac(X)$ gives the semistable bundle

$$L_1 \oplus L_2 \in M_X(2, 0).$$

This gives a map from the second symmetric product of $X \approx Jac(X)$ to $M_X(2, 0)$, a map which turns out to be an isomorphism.

Nevertheless, there are bundles on X which are not direct sums of line bundles. The group parametrizing extensions

$$0 \to L_1 \to V \to L_2 \to 0$$

is

$$Ext^1(L_2, L_1) \approx H^1(L_2^* \otimes L_1) \approx H^0(L_2 \otimes L_1^*)^*.$$

It is non-zero if and only if $L_1 \approx L_2$. So for every $L \in Jac(X)$ there is a non-trivial extension V of L by itself. This V is not isomorphic to $L_1 \oplus L_2$, but they are S-equivalent, so they occupy the same point in $\mathcal{M}_X(2,0)$.

Note that $\mathcal{M}_X(2,0)$ is 2-dimensional, while the dimension computed in Section 3.5.3 for the open subset $\mathcal{M}_X^s(2,0)$ was $(g-1)n^2 + 1 = 1$. In fact, there are no stable bundles, so the open subset is empty and there is no contradiction. Each semistable bundle has a 2-dimensional family of endomorphisms, so the dimension of the stack $\mathcal{M}_X(2,0)$ is $2 - 2 = 0$, in accordance with the general formula $(g-1)n^2$.

On the other hand, $\mathcal{M}_X(2,1)$ is 1-dimensional as it "should" be, and every semistable bundle of degree 1 is automatically stable. The situation for all ranks and degrees was worked out by Atiyah [At]: the dimension of $\mathcal{M}_X(n,d)$ is the greatest common divisor (n,d) of n and d; every S-equivalence class contains a unique representative which is the direct sum of (n,d) stable bundles, each of rank $n/(n,d)$ and degree $d/(n,d)$; and stable bundles exist if and only if $(n,d) = 1$, in which case all semistable bundles are stable and $\mathcal{M}_X(n,d)$ is 1-dimensional, as it should be. In general, the dimension of $\mathcal{M}_X(n,d)$ is (n,d) while the stack is always 0-dimensional. Finally, when the genus of X is $g \geq 2$, the open subset $\mathcal{M}_X^s(n,d)$ is always non-empty and dense in $\mathcal{M}_X(n,d)$, which has the predicted dimension $(g-1)n^2 + 1$.

3.5.6 Higgs bundles

A Higgs bundle is a pair (V, φ) where V is a vector bundle on X and

$$\varphi : V \to V \otimes K_X$$

is a 1-form valued endomorphism of V. A Higgs bundle (V, φ) is stable if the slope of every proper subbundle W *which is φ-invariant* is less than the slope of V; one defines semistable Higgs bundles similarly. Moduli spaces $\mathcal{H}iggs := \mathcal{H}iggs_X(n,d)$ and $\mathcal{H}iggs' \subset \mathcal{H}iggs$ exist, with properties analogous to those of $\mathcal{M}$ and $\mathcal{M}^s$. Note from our identification of cotangent vectors to $\mathcal{M}^s$ in Section 3.5.3 that there is a natural inclusion

$$T^* \mathcal{M}^s \subset \mathcal{H}iggs^s.$$

In fact for a stable V, the set of stable Higgs bundles (V, φ) can be identified with $T_V^* \mathcal{M}^s$. However, the stability condition for a Higgs bundle (V, φ) is weaker than that for the underlying vector bundle V, so $\mathcal{H}iggs^s$ is strictly bigger than $T^* \mathcal{M}^s$. Still, it can be checked that

$\mathcal{H}iggs^s$ is holomorphically symplectic: the symplectic form on $T^*\mathcal{M}^s$ extends to it and remains (closed and) non-degenerate.

Fix an effective divisor D on X, i.e. $D = \Sigma m_i p_i$ with $p_i \in X$, $m_i \geq 0$. A meromorphic Higgs bundle with poles on D is a pair (V, φ) where V is a vector bundle of rank n and degree d on X, and

$$\varphi : V \rightarrow V \otimes K_X(D)$$

is an endomorphism of V taking values in meromorphic differentials on X with pole on D. These have a moduli space $\mathcal{H}iggs_D := \mathcal{H}iggs_{X,D}(n, d)$ which is a key ingredient in the Theorem in section 3.6.1.

3.5.7 Other groups

The moduli spaces we have been considering have analogues for every reductive group G, the cases we have already seen corresponding to $G = GL(n)$. We go very briefly through some of the main points of the general case.

Instead of vector bundles, we consider principal G-bundles $\mathcal{V}$ on X. Given such a $\mathcal{V}$, each representation

$$\rho : G \rightarrow GL(n)$$

of G associates to $\mathcal{V}$ a vector bundle $\rho(\mathcal{V})$. The notions of stability and semistability can be defined directly for the principal bundles, in terms of the reductions of $\mathcal{V}$ to various parabolic subgroups of G. Given any representation ρ, we can also consider the (semi)stability of $\rho(\mathcal{V})$. Unfortunately, the (semi)stability of $\rho(\mathcal{V})$ may depend on the representation ρ. Fortunately, (semi)stability of $\rho(\mathcal{V})$ when ρ is in the adjoint representation turns out to be equivalent to the (semi)stability of the principal bundle. One obtains a projective moduli space $\mathcal{M} = \mathcal{M}_X^G$ parametrizing S-equivalence classes of semistable bundles, and an open subset $\mathcal{M}^s = \mathcal{M}_X^{s,G}$ parametrizing isomorphism classes of stable G-bundles, just as in the case $G = GL(n)$. There is also a moduli stack M_X^G parametrizing all G-bundles.

The tangent space to $\mathcal{M}_X^{s,G}$ at a non-singular point $\mathcal{V}$ is given, as in the case of $G = GL(n)$, by $H^1(X, ad(\mathcal{V}))$. For semisimple G,

$$dim(\mathcal{M}_X^G) = (g - 1) \cdot dim(G) \qquad (G \ semisimple)$$

since the generic G-bundle has no non-trivial automorphisms. For reductive groups there is a correction term equal to the dimension of the center of G, accounting for the automorphisms of stable bundles:

$$dim(\mathcal{M}_X^G) = (g-1) \cdot dim(G) + dim(Z(G)) \qquad (G \; reductive).$$

For the stack, the correction term drops out, so

$$dim(M_X^G) = (g-1) \cdot dim(G) \qquad (G \; reductive).$$

A G-Higgs bundle is a pair $(\mathcal{V}, \varphi)$ where $\mathcal{V}$ is a principal G-bundle and

$$\varphi \subset \Gamma(X, ad(\mathcal{V}) \otimes K_X),$$

and a meromorphic G-Higgs bundle with poles on an effective divisor D is a pair $(\mathcal{V}, \varphi)$ where now

$$\varphi \in \Gamma(ad(\mathcal{V}) \otimes K_X(D)).$$

The moduli spaces $\mathcal{H}iggs_{X,D}^G, \mathcal{H}iggs_{X,D}^{s,G}, \mathcal{H}iggs_X^G$ and $\mathcal{H}iggs_X^{s,G}$ exist and have the expected properties.

Finally note that, since we allow the group G to be non-semisimple, the resulting moduli spaces may be disconnected. The the components $\mathcal{M}_{X,d}^G$ and $\mathcal{H}iggs_{X,D,d}^G$ are indexed by the "degree" d of the G-bundle $\mathcal{V}$. This degree is a cocharacter of the center Z of G, i.e. d is an element of the lattice $Hom(\mathbf{C}^*, Z)$. When G is semisimple, its center is finite so there are no non-trivial cocharacters or components. For $G = GL(n)$ the cocharacter lattice is just the integers, so the "degree" is the usual degree, d, of the rank-n vector bundle V associated to the $GL(n)$-bundle $\mathcal{V}$. In this case we retrieve the moduli spaces $\mathcal{M}_X(n,d)$ of §5.2 and $\mathcal{H}iggs_{X,D}(n,d)$ of Section 3.5.6.

3.5.8 Other constructions

As for line bundles, we can ask which rank n vector bundles over a curve X admit a flat connection with structure group either $GL(n, \mathbf{C})$ or $U(n)$. A necessary condition is that the Chern class c_1 must vanish. The theorem of Narasimhan and Seshadri states that a rank n vector bundle with $c_1 = 0$ admits a flat connection if and only if it is *polystable*, i.e. the direct sum of stable bundles, each with $c_1 = 0$. In particular, each S-equivalence class of semistable bundles contains a unique representative which admits a flat $U(n)$ connection, so $\mathcal{M}_X(n, 0)$ can also be considered as the moduli space of flat $U(n)$ bundles on X. (This result has analogues, due to Donaldson, Uhlenbeck and Yau, valid over any compact Kähler X. There are also analogues involving $G_{compact}$ flat connections on polystable G-bundles, where G is the complexification of a compact group $G_{compact}$.)

The analogue of the Abel-Jacobi map involves the Hecke correspondences. A correspondence between two varieties is any subvariety of their product, considered as the graph of a "multivalued map" between them. Given a point $x \in X$, the ith Hecke correspondences $T^i \subset \mathcal{M} \times \mathcal{M}$ assigns to a vector bundle V the family of all vector bundles V' such that the sheaf of sections of V' contains that of V, with index i, and is contained in that of $V \otimes \mathcal{O}_X(x)$. (This family is parametrized by the Grassmannian of i-dimensional subspaces of the fiber V_x of V at x.) Starting with a given vector bundle V_0, any other vector bundle V can be reached by a sequence of Hecke correspondences (going both "up" and "down" if necessary). We will discuss neither flat bundles nor the Hecke correspondences any further in these notes.

3.6 Algebraic geometry of Higgs bundles

3.6.1 The theorem

Fix a curve X, an effective divisor D on X and a reductive group G. As in Section 3.5.7, we have a moduli space $\mathcal{M}_X^G$ parametrizing G-bundles $\mathcal{V}$ on X and a moduli space $\mathcal{Higgs}_{X,D}^G$ parametrizing meromorphic $G-$ $Higgs$ bundles (V, φ) with Higgs field $\varphi \in \Gamma(X, ad(V) \otimes K_X(D))$. As in Section 3.5.2, each of these moduli spaces comes in three flavors: spaces parametrizing stable objects or S-equivalence classes of semistable objects, and a stack parametrizing all objects. We are finally ready to state the central result:

Theorem
$\mathcal{Higgs}_{X,D}^G$ is the total space of an algebraically integrable system.

This holds in full generality for the stack version. Some minor restrictions (excluding curves X of low genus, or divisors D of low degree) are required in order to make this work for the moduli spaces of stable or semistable Higgs bundles, due to special features such as the existence of unexpected automorphisms, cf. Theorem (4.8) of [DM]. In these notes we will ignore these technical complications.

In order to prove the theorem, we need to exhibit on $\mathcal{Higgs}_{X,D}^G$ an algebraic Poisson structure ψ and an algebraic map $H : \mathcal{Higgs}_{X,D}^G \to B_{X,D}^G$ which is Hamiltonian with respect to ψ and whose generic fibers, or 'Liouville tori', are abelian varieties. The symplectic foliation of

$(\mathcal{Higgs}^G_{X,D}, \psi)$ should be pulled back from an algebraic foliation of $B^G_{X,D}$. In fact, we will find another space $C^G_{X,D}$ and an algebraic map

$$B^G_{X,D} \to C^G_{X,D}$$

whose fibers give the generic leaves in $B^G_{X,D}$: the Casimirs on $\mathcal{Higgs}^G_{X,D}$ are the pullbacks of functions on $C^G_{X,D}$, while the larger collection of all Hamiltonians consists of the pullbacks of functions on $B^G_{X,D}$.

In the remainder of Section 3.6 we discuss the algebro-geometric aspects of the theorem: we construct the Hamiltonians and Casimirs, and identify the Liouville tori as abelian varieties. The symplectic aspect, i.e. the construction of ψ, will be taken up in Section 3.7.

3.6.2 Spectral data and cameral covers

In this section we present a heuristic description of the Hamiltonian map and the Liouville tori. We start with the case $G = GL(n)$. We want to think of the Higgs bundle (V, φ) as consisting of two distinct pieces: its eigenvalues and eigenspaces. Very roughly, the Hamiltonians fix the eigenvalues while a point in the Liouville torus determines the eigenspaces.

An $n \times n$ matrix is regular semisimple if it is diagonalizable with distinct eigenvalues. Specifying such a matrix is equivalent to specifying n distinct eigenvalues plus the eigenline associated to each.

For a pair (V, φ) where V is a rank n vector bundle on X and $\varphi : V \to V$ is an everywhere regular semisimple endomorphism, the n moving eigenvalues determine an n-sheeted cover $\pi : \bar{X} \to X$ contained in $X \times \mathbf{C}$, while the moving eigenlines form a line bundle $L \in Pic(\bar{X})$. For an (everywhere regular semisimple) Higgs bundle $(V, \varphi : V \to V \otimes K_X(D))$, the only change is that $\bar{X}$ sits in the total space $Tot(K_X(D))$, rather than in $X \times \mathbf{C}$ which is the total space of the trivial bundle $\mathcal{O}_X$.

But the requirement that φ be everywhere regular semisimple is unrealistic. For generic (V, φ) we expect φ to be regular semisimple at all but a finite number of points of X. In this situation, the cover $\bar{X}$ becomes a branched cover $\pi : \bar{X} \to X$, still contained in $Tot(K_X(D))$. In general, $\bar{X}$ could be any subscheme of $Tot(K_X(D))$ whose projection to X is finite of degree n. When $\bar{X}$ is singular or non-reduced, the line bundle L can degenerate to various other sheaves on $\bar{X}$: the eigenspaces need not be eigenlines. But we can still recover the underlying vector bundle V via the direct image construction encountered in Section 3.5.5: $V = \pi_* L$.

So in conclusion, in case $G = GL(n)$, we can let $B_{X,D}^{GL(n)}$ be the space of all such "spectral covers" $\bar{X}$. The Hamiltonian map

$$H : \mathcal{H}iggs_{X,D}^{GL(n)} \to B_{X,D}^{GL(n)}$$

then sends (V, φ) to its spectral cover $\bar{X}$. The Liouville torus then parametrizes all the allowable sheaves L on $\bar{X}$. In the generic situation $\bar{X}$ will be non-singular, the sheaves L are just line bundles, and the Liouville torus is a component of $Pic(\bar{X})$. (As noted in Section 5.5, the degree of L is not the same as the degree of V. Rather, there is a shift: it is the Euler characteristic which is preserved.)

The generalization to arbitrary reductive G is very important and very pretty, but it requires a recasting of the simple eigenvalue/eigenvector picture. In the case $G = GL(n)$ this amounts to replacing the degree n spectral cover $\pi : \bar{X} \to X$ by a degree $n!$ *cameral cover* $\widetilde{X} \to X$. Over a point $x \in X$ where $\varphi(x)$ is regular semisimple, so $\pi^{-1}(x)$ consists of n distinct points $\{x_1, ..., x_n\}$, the points of $\widetilde{X}$ correspond to the $n!$ ways of ordering the x_i. $\widetilde{X}$ has the advantage that it is a Galois cover of X, with Galois group $\mathcal{S}_n$, the symmetric group. The n projections $\pi_i : \widetilde{X} \to \bar{X}$ give the n line bundles $L_i := \pi_i^* L$ on $\widetilde{X}$. Let $T \approx (\mathbf{C}^*)^n$ be the maximal torus of $GL(n)$, i.e the subgroup of diagonal matrices. The decomposable rank n vector bundle $\oplus_{i=1}^n L_i$ determines a principal T-bundle $\mathcal{T}$ and we can work out how $\mathcal{T}$ transforms under the action of $\mathcal{S}_n$. Our recasting amounts to replacing the pair $(\bar{X}, L)$ by $(\widetilde{X}, \mathcal{T})$, where $\widetilde{X}$ is $\mathcal{S}_n$-Galois over X and $\mathcal{T}$ is a T-bundle with the correct transformation property.

This seems more complicated, but it has the advantage that it extends to all groups G. The symmetric group $\mathcal{S}_n$ is replaced by the Weyl group W of G, $W = N_G(T)/T$ where $N_G(T)$ is the normalizer of T in G. The $n!$ points in $\widetilde{X}$ over a point $x \in X$ for which $\varphi(x)$ is regular semisimple are replaced by the collection of *chambers* in the Cartan subalgebra containing φ. The cover $\widetilde{X}$ is therefore called the *cameral* cover. More generally, as long as $\varphi(x)$ is regular (but not necessarily semisimple), the fiber of $\widetilde{X}$ over $x \in X$ can be identified with the set of Borel subalgebras in the Lie algebra $ad(V_X)$ containing the element $\varphi(x) \otimes \alpha^{-1}$, for any non-zero element α in the fiber $(K_X(D))_{|x}$ of the line bundle $K_X(D)$ at the point x. Thus a G-Higgs bundle (V, φ) determines a cameral cover $\widetilde{X} \to X$ together with a principal B-bundle on $\widetilde{X}$, where B is a Borel subgroup of G. Since the maximal torus T is recovered from B as $T = B/[B, B]$, there is an associated T-bundle on $\widetilde{X}$ which we denote $\mathcal{T}$. The bundles $\mathcal{T}$ which arise this way transform under the Weyl group

W according to an affine transformation law which is worked out in general in [D1] and [DG]. It is a shifted form of the natural action of W on T-bundles, which is induced from the action of W on the lattice Λ of characters of G, but the affine shift is quite delicate. The bottom line is that there is a categorical equivalence between regular G-Higgs bundles $(\mathcal{V}, \varphi)$ and pairs $(\widetilde{X}, \mathcal{T})$ consisting of a cameral cover and a properly-transforming T-bundle on it. We let $B_{X,D}^G$ be a parameter space for cameral covers. The Hamiltonian map $H : \mathcal{H}iggs_{X,D}^G \to B_{X,D}^G$ sends a G-Higgs bundle $(\mathcal{V}, \varphi)$ to the point of $B_{X,D}^G$ parametrizing the corresponding cameral cover $\widetilde{X}$. The family of *all* T-bundles on $\widetilde{X}$ is the product of n copies of $Pic(\widetilde{X})$ where n is the rank of G, i.e. the dimension of T or the rank of the lattice Λ. Or, more intrinsically, this family is given by $Hom(\Lambda, Pic(\widetilde{X}))$. This is an abelian variety, as is its subgroup $Hom_W(\Lambda, Pic(\widetilde{X}))$ of W-equivariant T-bundles, called the *generalized Prym variety*. The Liouville torus is a nontrivial coset of this subgroup (reflecting the fact that the action is affine rather than linear), so it is also isomorphic to an abelian variety.

It is worth noting that each representation

$$\rho : G \to GL(N)$$

converts a G-Higgs bundle to its associated $GL(n)$-Higgs bundle. The latter has an N-sheeted spectral cover $\bar{X}_\rho$, depending on ρ. Hitchin's original approach in [H], in case $D = 0$ and G a classical group, was based on these spectral covers for the "classical representations" of G. The $\bar{X}_\rho$ for all representations ρ are recovered as associated covers of these cameral covers $\widetilde{X}$.

3.6.3 Hamiltonians and Casimirs

We want to write down explicit expressions for the Hamiltonian base $B_{X,D}^G$, the Casimir base $C_{X,D}^G$, and the maps to them.

For $G = GL(n)$, recall that the spectral cover $\bar{X}$ is an arbitrary sub-scheme of $Tot(K_X(D))$ which is finite of degree n over X. In terms of a local coordinate x on X and a linear coordinate y along the fibers of $K_X(D)$, such an $\bar{X}$ is given by a polynomial equation

$$f(x, y) = y^n + b_1(x)y^{n-1} + \cdots + b_n(x) = 0.$$

This still works globally, but the b_i are not functions – they are sections of line bundles $(K_X(D))^{\otimes i}$. In fact if $\bar{X}$ is the spectral cover of $(\mathcal{V}, \varphi)$, then $f(x, y)$ is just the characteristic polynomial of φ, so

$$b_i(x) = s_i(\varphi)$$

where s_i is the ith symmetric function in the roots of φ. So $B_{X,D}^{GL(n)}$, which is the space parametrizing all equations $f(x, y) = 0$, is the sum of the spaces parametrizing the individual coefficients:

$$B_{X,D}^{GL(n)} = \bigoplus_{i=1}^{n} \Gamma\Big(X, (K_X(D))^{\otimes i}\Big),$$

and

$$H : \mathcal{H}iggs_{X,D}^{GL(n)} \to B_{X,D}^{GL(n)}$$

sends

$$(\mathcal{V}, \varphi) \mapsto (s_1(\varphi), ..., s_n(\varphi)).$$

We can also describe the quotient $C_{X,D}^{GL(n)}$, i.e. specify the subset of the Hamiltonian functions which will turn out (once we defined ψ) to be the Casimirs. It is

$$C_{X,D}^{GL(n)} = \bigoplus_{i=1}^{n} \Gamma\Big(D, ((K_X(D))^{\otimes i})_{|D}\Big),$$

and the map

$$Res : B_{X,D}^{GL(n)} \longrightarrow C_{X,D}^{GL(n)}$$

simply sends the n-tuple (b_i) of sections of line bundles on X to the n-tuple of their restrictions to the divisor D. The composition

$$\mathcal{H}iggs_{X,D}^{GL(n)} \longrightarrow B_{X,D}^{GL(n)} \longrightarrow C_{X,D}^{GL(n)}$$

then sends a Higgs bundle $(\mathcal{V}, \varphi)$ to the n-tuple of symmetric functions of the residue $res_D(\varphi)$ which is a "Higgs field" on the 0-dimensional scheme D, with values in the canonical line bundle $(K_X(D))_{|D} \approx K_D$ of D itself:

$$
\begin{array}{ccc}
\mathcal{H}iggs_{X,D}^{GL(n)} & \xrightarrow{H_{X,D}^G} & B_{X,D}^{GL(n)} \\
res = \downarrow residue & & Res = \downarrow Restriction \\
\mathcal{H}iggs_D^{GL(n)} & \xrightarrow{H_D^G} & C_{X,D}^{GL(n)}.
\end{array}
$$

For a reductive group G of rank n there are n polynomial functions $s_i^G, i = 1, ..., n$ on the Cartan subalgebra $\mathbf{t}$ of G (i.e. the Lie algebra of the maximal torus T), which generate the algebra of W-invariant polynomials on $\mathbf{t}$. For $G = GL(n)$, these are the elementary symmetric functions, so the degree of $s_i^{GL(n)}$ is i. In general, the degrees $d_i = deg(s_i^G)$ are invariants of the group G, e.g. for $G = SO(2n+1)$ we have $d_i = 2i$, while for $G = SO(2n)$ we have $d_i = 2i$ for $i < n$, and $d_n = n$. (The corresponding invariant function $s_n^{SO(2n)}$ is the Pfaffian – the square root of the determinant, which is a well defined $SO(2n)$-invariant function of a skew symmetric matrix.)

It turns out in general that fixing a cameral cover is equivalent to fixing the values of these n invariant polynomials. The Hamiltonian base is therefore:

$$B_{X,D}^G = \bigoplus_{i=1}^n \Gamma\Big(X, (K_X(D))^{\otimes d_i}\Big),$$

and the Hamiltonian

$$H : \mathcal{Higgs}_{X,D}^G \longrightarrow B_{X,D}^G$$

sends

$$(V, \varphi) \mapsto \Big(s_i^G(\varphi)\Big)_{i=1}^n.$$

The Casimir base $C_{X,D}^G$ is obtained similarly, by restricting from X to D.

3.7 Symplectic geometry of Higgs bundles

3.7.1 Moment maps

Recall from Section 3.2.4 that a Poisson structure ψ on a manifold M determines a Lie algebra homomorphism $v : C^\infty(M) \to VF(M)$. Vector fields in the image of v, i.e. of the form $v(f)$ for a C^∞ function f on M, are called Hamiltonian. A vector field X is locally Hamiltonian if there is an open covering $\{\mathcal{U}_i\}$ of M such that the restriction of X to each $\mathcal{U}_i$ is Hamiltonian.

An action $\rho : G \times M \to M$ of a connected Lie Group G (with Lie algebra $\mathbf{g}$) on a manifold M determines an infinitesimal action $d\rho$ which is a Lie algebra homomorphism:

$$d\rho : \mathbf{g} \to VF(M).$$

Conversely, ρ can be recovered by exponentiating $d\rho$.

When (M, φ) is a Poisson manifold we say that the action ρ is Poisson (or: liftable) if there is a Lie algebra homomorphism $\ell : \mathbf{g} \to C^\infty(M)$ which lifts $d\rho$, i.e. such that $v \circ \ell = d\rho$. In particular, the image of $d\rho$

consists of Hamiltonian vector fields. If the action ρ is Poisson, then for every $g \in G$ the diffeomorphism $\rho(g) : M \to M$ preserves the Poisson structure ψ. In fact, the latter condition is equivalent to the image of $d\rho$ consisting of locally Hamiltonian vector fields.

The lift ℓ determines (and is equivalent to) a map

$$\mu : M \to \mathbf{g}^*$$

defined by

$$< \mu(m), A >:= \ell(A)(m).$$

This map μ is called the *moment map* for the Poisson action ρ. It is a Poisson map, i.e. it sends the Poisson structure ψ on M to the Kirillov-Kostant Poisson structure on $\mathbf{g}^*$. It is also G-equivariant. We illustrate this with the two examples from Section 3.2:

(i) Any action ρ of G on a manifold Y induces an action $T^*\rho$ of G on the symplectic $M := T^*Y$. This action $T^*\rho$ is Poisson. Indeed the infinitesimal action of ρ

$$d\rho : \mathbf{g} \to VF(Y) = \Gamma(Y, TY)$$

composes with the evaluation map

$$eval : \Gamma(Y, TY) \longrightarrow C^\infty(T^*Y)$$

to yield the desired lift

$$\ell : \mathbf{g} \longrightarrow C^\infty(T^*Y).$$

The corresponding moment map

$$\mu : T^*Y \longrightarrow \mathbf{g}^*$$

is the fiber by fiber dual of $d\rho$. It can also be interpreted as the pullback, via ρ, of differential forms from Y to G.

(ii) The coadjoint action of G on the Poisson manifold $\mathbf{g}^*$ is also a Poisson action. Its moment map is the identity

$$\mathbf{g}^* \to \mathbf{g}^*.$$

3.7.2 Symplectic reduction

Let ρ be a Poisson action of a connected Lie group G on a symplectic manifold M, ω. Assume the action is nice enough that the quotient M/G

is a manifold. We then have the quotient map

$$M \to M/G$$

and the moment map

$$\mu : M \to \mathbf{g}^*.$$

If G is abelian, the G-equivariance of μ means that it factors through M/G:

$$\mu : M \to M/G \to \mathbf{g}^*$$

and for each $\xi \in \mathbf{g}^*$, the subquotient

$$M_\xi := \mu^{-1}(\xi)/G$$

is again a symplectic manifold, called the symplectic reduction of M at ξ.

When G is non-abelian, μ does not factor through M/G. Instead we consider the coadjoint orbit $\mathcal{O}_\xi \subset \mathbf{g}^*$ which is the symplectic leaf in $\mathbf{g}^*$ containing ξ, and set

$$M_\xi := \mu^{-1}(\mathcal{O}_\xi)/G \approx \mu^{-1}(\xi)/G(\xi),$$

where G_ξ is the stabilizer of ξ in G. Again, M_ξ inherits a symplectic structure from M, and is called the symplectic reduction of M at ξ.

In fact, all these symplectic reductions fit together. The invertible Poisson structure ω^{-1} on M is immediately seen to descend to a Poisson structure on the quotient M/G. This is no longer symplectic; its symplectic leaves are precisely the symplectic reductions M_ξ, so they are parametrized by the coadjoint orbits $\mathcal{O}_\xi$ in $\mathbf{g}^*$.

3.7.3 The Poisson structure on $\mathcal{H}iggs_{X,D}^G$

We want to obtain the Poisson structure on $\mathcal{H}iggs_{X,D}^G$ by symplectic reduction. The idea is as follows. Let a group G act on a manifold Y with a nice quotient $Z = Y/G$. The induced action ρ of G on the cotangent bundle $M := T^*Y$ is automatically Poisson, as we saw in Example (1) of Section 3.7.1. So, as in Section 3.7.2, the symplectic structure on T^*Y descends to a Poisson structure on $W := T^*Y/G$. If G acts freely on Y, this W is a vector bundle over Z: if $y \in Y$ maps to $z \in Z$ then the fiber of W over z can be identified with T_y^*Y.

We want to rig things so that $Z = \mathcal{M}_X^G$, $W = \mathcal{H}iggs_{X,D}^G$, and the map $W \to Z$ sends a G-Higgs bundle $(\mathcal{V}, \varphi)$ to the underlying G-bundle $\mathcal{V}$.

There are various complications and technicalities. For instance, there are (semi)stable Higgs bundles $(\mathcal{V}, \varphi)$ involving an unstable $\mathcal{V}$, so the "map" $W \to Z$ is not everywhere defined (unless we work with stacks). A related problem is that some of the objects have extra automorphisms, causing the action of G not to be very nice. These details are handled in [M], [DM] and elsewhere. So here I will ignore them. This produces a Poisson structure only on a Zariski open subset of $\mathcal{Higgs}^G_{X,D}$ and extra work is needed in order to extend it, cf. [DM], Section 3.5.4.

So we are searching for a space Y with an action of a group G such that Y/G is (generically) $\mathcal{M}^G_X$, while a typical cotangent space can be identified with the fiber of $\mathcal{Higgs}^G_{X,D}$ over $\mathcal{M}^G_X$:

$$T^*_y Y \approx \Gamma(X, ad(\mathcal{V}) \otimes K_X(D))$$

for $y \in Y$ above $\mathcal{V} \in \mathcal{M}^G_X$.

Recall the deformation theory worked out in Section 3.5.3 for $GL(n)$-bundles and in Section 3.5.7 for arbitrary G-bundles. We have identifications

$$T_{\mathcal{V}} \mathcal{M}^G_X \approx H^1(X, ad(\mathcal{V}))$$

and dually

$$T^*_{\mathcal{V}} \mathcal{M}^G_X \approx \Gamma(X, ad(\mathcal{V}) \otimes K_X).$$

Compare with what we want of Y:

$$T^*_y Y \approx \Gamma(X, ad(\mathcal{V}) \otimes K_X(D)),$$

or dually:

$$T_y Y \approx H^1(X, ad(\mathcal{V}) \otimes \mathcal{O}_X(-D)).$$

So while a deformation of a G-bundle $\mathcal{V} \in \mathcal{M}^G_X$ is given by an $ad(\mathcal{V})$-valued 1-cocycle $\{h_{ij}\}$, a deformation of a point of Y should be given by a cocycle $\{h_{ij}\}$ *which vanishes on D.*

This suggests the correct choice: for Y we take the moduli space $\mathcal{M}^G_{X,D}$ of G-bundles on X together with a level D structure. By definition a level-D structure on a G-bundle $\mathcal{V}$ is a trivialization of $\mathcal{V}|_D$, i.e. an isomorphism

$$\eta : \mathcal{V}|_D \xrightarrow{\sim} G(\mathcal{O}_D)$$

where $G(\mathcal{O}_D)$ is the trivial G-bundle on D. A deformation of the pair $(\mathcal{V}, \eta)$ is given by a 1-cocycle $\{g_{ij}\}$ which is compatible with the trivialization η; this means precisely that the additive cocycle $\{h_{ij} = g_{ij}^{-1} g'_{ij}\}$ vanishes on D, as required.

For the group acting on $\mathcal{M}_{X,D}^G$ with quotient $\mathcal{M}_X^G$ we can take

$$G_D' := Aut(G(\mathcal{O}_D)).$$

When D consists of d distinct points, an element of G_D' is a d-tuple of elements of G. An element $g \in G_D'$ acts on $\mathcal{M}_{X,D}^G$ sending $(\mathcal{V}, \eta)$ to $(\mathcal{V}, g \circ \eta)$. On the other hand, if α is an automorphism of $\mathcal{V}$, then the pairs $(\mathcal{V}, \eta)$ and $(\mathcal{V}, \eta \circ \alpha)$ represent the same point of $\mathcal{M}_{X,D}^G$. In particular, every element z of the center $Z(G)$ of G determines such an automorphism, $\alpha(z)$. The action of G_D' on $\mathcal{M}_{X,D}^G$ therefore factors through an action of the quotient group:

$$G_D := Aut(G(\mathcal{O}_D))/Z(G).$$

This quotient group acts generically freely on $\mathcal{M}_{X,D}^G$, with quotient $\mathcal{M}_X^G$. As usual, this action lifts to $T^*\mathcal{M}_{X,D}^G$, and the quotient $T^*\mathcal{M}_{X,D}^G/G_D$ can be identified (away from a bad locus) with $\mathcal{H}iggs_{X,D}^G$, which thus has a natural Poisson structure. The symplectic leaves are precisely the fibers over the Casimir base, which has a natural interpretation via the coadjoint action of G_D:

$$C_{X,D}^G \approx (\mathbf{g}_D)^*/G_D.$$

3.8 Examples, further developments, open problems

3.8.1 Some examples

The abelian case

The case $G = GL(1)$ is quite trivial: the moduli space of bundles $\mathcal{M}G_X$ is $Pic(X)$, the moduli space of Higgs bundles $\mathcal{H}iggs_{X,D}^G$ is the product $Pic(X) \times \Gamma(X, K_X(D))$, the Hamiltonian base is $B_{X,D}^G = \Gamma(X, K_X(D))$, and the Hamiltonian map is the projection to the second factor. The Casimir base is $C_{X,D}^G = \Gamma(D, K_X(D)|_D)$, and each symplectic leaf in $\mathcal{H}iggs_{X,D}^G$ is isomorphic to the cotangent bundle $T^*Pic(X) = Pic(X) \times \Gamma(X, K_X)$.

Hitchin's system

The case $D = 0$ of *holomorphic* Higgs bundles was studied by Hitchin [H]. As we saw in Section 3.5, the total space $\mathcal{H}iggs_X^G$ is a partial compactification of the cotangent bundle $T^*\mathcal{M}G_X$ to moduli. Hitchin wrote down the Hamiltonians (i.e. the coefficients of the characteristic polynomial) in this case, and proved the complete integrability when the group G is one of the classical groups $GL(n), SL(n), SO(n), Sp(n)$. Proofs of several versions of the integrability for general G can be found in [BK, D1, DG, F, S].

Rational base

In case $X = \mathbf{P}^1$ and $G = GL(n)$, the system of meromorphic Higgs bundles was described explicitly in [AHH, Be]. As we saw in Section 3.5.5, the stack $M_{\mathbf{P}^1}(n, 0)$ is stratified: it has one dense point, corresponding to the trivial bundle $\mathcal{O}_X^{\oplus n}$, plus an infinite nesting of other points in its closure. The forgetful map $\mathcal{H}iggs_{\mathbf{P}^1, D}^G \to M_{\mathbf{P}^1}(n, 0)$ induces a stratification of $\mathcal{H}iggs_{\mathbf{P}^1, D}^G$, with a dense open subset $(\mathcal{H}iggs_{\mathbf{P}^1, D}^G)^0$ parametrizing Higgs bundles whose underlying vector bundle is the trivial bundle. If D happens to be d times the point p_∞ where a coordinate t on $\mathbf{P}^1$ becomes infinite, a Higgs bundle in the open subset $(\mathcal{H}iggs_{\mathbf{P}^1, d \cdot p_\infty}^G)^0$ can be given by a polynomial $\sum_{i=0}^{d-2} A_i t^i \, dt$ whose coefficients A_i are $n \times n$ matrices. If D is the sum of d distinct points p_i where $t = t_i$, the Higgs bundle in the open subset $(\mathcal{H}iggs_{\mathbf{P}^1, d \cdot p_\infty}^G)^0$ can be given instead by a rational function $\sum_{i=1}^d B_i / (t - t_i) \, dt$, with $n \times n$ matrices B_i satisfying $\sum_{i=1}^d B_i = 0$. Markman's Poisson structure can be written explicitly in terms of the A's or B's. The special case of Theorem 3.6.1 proved in [AHH, Be] says that the set of such polynomial (or rational) matrices with fixed characteristic polynomial (equivalently, spectral cover $\overline{C}$) is an open subset of the Jacobian of $\overline{C}$. The complement, coming from Higgs bundles whose underlying vector bundle is not trivial, can also be described explicitly in terms of the theta divisor of $Jac(\overline{C})$.

The particular case of $G = SL(2)$ was studied earlier by Mumford [Mum], and used in his solution of the Schottky problem for hyperelliptic curves. Particular choices of the residues lead to very classical systems such as the geodesic flow on an ellipsoid and Neumann's system, cf. [Be, DM].

Elliptic base

A similarly explicit description is available when the base X has genus $g = 1$, using the description of the moduli space given in Section 3.5.5. A particularly important case arises when D is one point p_∞ and the residue there is in the conjugacy class of the diagonal matrix $O := diag(1, \ldots, 1, 1 - n)$. The spectral curves which arise are the *elliptic solitons* studied by Krichever [Kr] and Treibich–Verdier [TV]. These are n-sheeted covers $\pi : C \to X$, where C has genus n and at least $n-1$ of the n sheets come together above p_∞. Equivalently, $Jac(C)$ contains a copy of the elliptic curve X, and the image $AJ(C)$ of the Abel-Jacobi map (cf. Section 3.4.4) is *tangent* to X at the ramification point. (According to Krichever, the Jacobian $Jac(C)$ together with the linear flows on it given by the derivatives of AJ constitute a KP soliton, i.e. a finite-dimensional orbit of the infinite-dimensional *KP hierarchy*. An elliptic soliton is such a solution in which the first KP flow is tangent to an elliptic curve, hence is doubly periodic.)

The same system arises, from very different considerations, in supersymmetric Yang-Mills theory, cf. [DW]. The relation of integrable systems to Seiberg–Witten theory and supersymmetric Yang-Mills is discussed in [D2], where various related spectral curves are worked out explicitly. This elliptic soliton system is also the $SL(n)$ case of the *elliptic Calogero–Moser system*, which suggests an analogue for other groups, cf. [DHP].

3.8.2 Further developments

Bundles on elliptic fibrations

We noted in Section 3.6.2 that there is a categorical equivalence between regular G-Higgs bundles $(\mathcal{V}, \varphi)$ and pairs $(\widetilde{X}, \mathcal{T})$ consisting of a cameral cover and a properly-transforming T-bundle on it. The version we needed there involved, on the one hand, $K_X(D)$-valued G-Higgs bundles, and on the other hand, cameral covers $\widetilde{X}$ mapped to the total space $Tot(K_X(D))$: this allowed us to have parameter spaces $B^G_{X,D}$, for the cameral covers, and $\mathcal{H}iggs^G_{X,D}$, for the Higgs bundles, with a map $\mathcal{H}iggs^G_{X,D} \to B^G_{X,D}$ whose fibers are the generalized Prym varieties $Hom_W(\Lambda, Pic(\widetilde{X}))$. However, the result of [D1] and [DG] is stronger: it is a categorical equivalence between *abstract* G-Higgs bundles (with no specification of the bundle $K_X(D)$ in which they take their *values*), and abstract spectral data $(\widetilde{X}, \mathcal{T})$, consisting of an abstract cameral cover $\widetilde{X}$ (with no specification of a map to the total space of a bundle) plus a properly-transforming T-bundle on it. (The definition of an abstract G-Higgs bundle is somewhat technical: it is defined [D1, Def. 7] to be a pair $(\mathcal{V}, \mathbf{c})$, where $\mathcal{V}$ is, as usual, a principal G-bundle on X, and $\mathbf{c} \subset ad(\mathcal{V})$ is a subbundle which fiber by fiber consists of *regular centralizers*, i.e. centralizers in $\mathbf{g}$ of regular elements of G.) The "valued" result follows immediately from the "abstract" one by adding to the Higgs bundle $(\mathcal{V}, \mathbf{c})$ a section $\varphi \in \Gamma(X, \mathbf{c} \otimes K_X(D))$, and to the spectral data $(\widetilde{X}, \mathcal{T})$ an appropriate map $\widetilde{X} \to Tot(K_X(D))$. (Actually, this should be a W-equivariant collection of maps, cf. [D1, Def. 6].)

But given the notions of abstract Higgs bundles and spectral data, we can clearly consider Higgs bundles and spectral data with values in any family Y of groups over X, not necessarily in a line bundle: to a an abstract Higgs bundle add a section $\varphi \in \Gamma(X, \mathbf{c} \otimes Y)$, and to the abstract spectra data $(\widetilde{X}, \mathcal{T})$ add an appropriate collection of maps to $Tot(Y)$. With the obvious notation, we get parameter spaces $\mathcal{H}iggs^G_{Y/X}$ and $B^G_{Y/X}$ and a map $\mathcal{H}iggs^G_{Y/X} \to B^G_{Y/X}$ whose fibers are the generalized Prym varieties $Hom_W(\Lambda, Pic(\widetilde{X}))$.

An important case involves "values" in an elliptic fibration, i.e. a variety Y fibered over X with the generic fiber being an elliptic curve. This is important due to the observation [D3] that the moduli space $\mathcal{M}^G_Y$ of G-bundles on Y (satisfying an appropriate version of stability) can be identified with the moduli space $\mathcal{H}iggs^G_{Y/X}$ of G-Higgs bundles on X with values in the fibration $Y \to X$. Combining this with the equivalence of Higgs bundles with spectral data, we see that $\mathcal{M}^G_Y$ can be fibered over a base $B^G_{Y/X}$, the fibers being the same generalized Prym varieties $Hom_W(\Lambda, Pic(\widetilde{X}))$ which occur for line-bundle-valued Higgs bundles.

In case Y is a 2-dimensional algebraically integrable system, fibered over a curve X with elliptic fibers, the resulting $\mathcal{M}^G_Y$ is an algebraically integrable system with base $B^G_{Y/X}$ and generalized Prym varieties for fibers. (The case that $G = GL(n)$ is also a special case of Mukai's system, see below.) When $dim(Y) \geq 3$ we lose the symplectic aspects of an algebraically integrable system, and retain only the fibration by abelian varieties. The point is that the moduli space of bundles on Y can be analyzed in terms of objects on the lower dimensional base X.

Another approach to the study of $\mathcal{M}^G_Y$ is in [FMW]. This is based on Looijenga's result [Lo] which says that the moduli of G bundles on an elliptic curve, for any semisimple group G, is a weighted projective space, and on the use of *minimally unstable* bundles and their deformations.

Non-linear deformations

Mukai [Mu] showed that on an algebraically symplectic surface Y, each component of the moduli space of coherent sheaves is itself algebraically symplectic (away from its possible singularities). We have already encountered the case of vector bundles on an elliptically fibered symplectic surface. In order to obtain another important case, choose a curve $\overline{C}$ with an embedding $i : \overline{C} \to Y$ and a line bundle on it, $L \in Pic(\overline{C})$, and consider the moduli space $Muk_{Y, \overline{C}, L}$ of coherent sheaves on Y whose Hilbert polynomial equals that of the direct image $i_*(L)$. The support of such a sheaf is a curve algebraically equivalent to $\overline{C}$, and there is a well-defined support map $H : Muk_{Y, \overline{C}, L} \to |\overline{C}|$ to the moduli space (often, the linear system) $|\overline{C}|$ of curves in Y algebraically equivalent to $\overline{C}$. This support map is Lagrangian with respect to the symplectic structure on $Muk_{Y,\overline{C},L}$, and the fiber over a non-singular $\overline{C}$ is a component of $Pic(\overline{C})$.

In the case when $Y = T^*X$ and $\overline{C} = n \cdot X$ is n times the zero section, $Muk_{T^*X, n \cdot X, L}$ is just Hitchin's system. Note that although Y in this case is not compact, the linear system $|\overline{C} = n \cdot X|$ is finite dimensional, in fact it is precisely the system of Hitchin spectral curves. (More generally, Tyurin [T] extended Mukai's result to Poisson surfaces Y. Theorem 6.1 is obtained by taking $Y := Tot(K_X(D))$.)

On the other hand, we can take X to be any non-singular curve contained in a K3 surface Y. It is shown in [DEL] that $Muk_{Y, n \cdot X, L}$ is a non-linear deformation of $Higgs_X^{GL(n)}$. The fiber over the non-reduced divisor $n \cdot X$ is seen to be a certain affine twist of the nilpotent cone in Hitchin's system, analyzed by Laumon [La]. In particular, this fiber is reducible: one component parametrizes vector bundles on the reduced X, while other components parametrize sheaves on Y whose support is the full non-reduced $\overline{C} = n \cdot X$, for example line bundles in $Pic(n \cdot X)$. This analysis seems to be useful for understanding the D-branes of nonperturbative string theory, cf. [GS] for a recent example.

3.9 Open problems

Seiberg-Witten integrable systems
Quantum field theoretic considerations [SW1,2] predict the existence of a collection of *Seiberg-Witten integrable systems*. These should depend on the choice of an elliptic curve X plus a complex semisimple group G and a collection of representations of G satisfying certain numerical conditions, cf. Section 1 of [D2]. The Liouville tori for these systems should be r-dimensional, where r is the rank of G, and they should be Jacobians or Prym varieties of the Seiberg-Witten curves, which play the role of spectral curves for these systems. In particular, there should exist a Seiberg-Witten integrable system (SWIS) for each elliptic curve X and complex semisimple group G taken with its *adjoint* representation. The case $G = SL(n)$ was solved in [DW]. The system obtained there is equivalent to the elliptic solitons system. It is tempting to guess that the SWIS for any group G (with its adjoint representation) is similarly obtained from the meromorphic Higgs bundles on X with structure group G. The problem is that the dimension of the Liouville tori for such a system equals half the dimension of the coadjoint orbit to which the

residue is confined; for groups other than $SL(n)$, a small enough orbit (i.e. of dimension $2r$) does not seem to exist.

D'Hoker and Phong have shown [DHP] that integrable systems satisfying the SW requirements can be obtained by choosing appropriate parameters in the *elliptic Calogero-Moser system* for the group G. However, these choices are made on a case-by-case basis. It would be very interesting to understand a priori which choices work, and to derive the entire collection of SWIS simultaneously from some uniform geometric construction, presumably some variant of the moduli space of meromorphic Higgs bundles.

Mathematics of string duality

The central development in string theory in the last few years has been the discovery of *string dualities*. The various individual perturbative string theories are now believed to be equivalent to each other and to represent different limits of a single *M-theory*. The two central dualities are the one between types IIA and IIB strings, known as *mirror symmetry*, and the duality between the heterotic string and F-theory, cf. [V, MV]. Among other things, this conjectured duality predicts a natural isomorphism between the moduli spaces of the two theories. The heterotic moduli space parametrizes collections of data which include, among other fields, a Calabi–Yau n-fold Z together with a G-bundle on Z, where G is $E8 \times E8$ (where $E8$ is the 248-dimensional exceptional Lie group) or a subgroup in it. The corresponding F-theory moduli space parametrizes data which includes, among other fields, an elliptically fibered Calabi-Yau $n + 1$-fold $Y \to X$, and a "Ramond–Ramond" field, which amounts to a point in the *Deligne cohomology group* of Y, a certain extension of the intermediate Jacobian $Jac(Y)$ of Clemens and Griffiths [CG]. The best understood case [MV] is when Z is itself elliptically fibered over a base B. X is then a $\mathbf{P}^1$-bundle over the same B, and Y is fibered over B with K3 fibers (and these K3's in turn are elliptically fibered over the $\mathbf{P}^1$'s). In the limit when one of the auxiliary fields on the heterotic side goes to zero, the F-theory varieties degenerate into reducible varieties $X = X_1 \cup X_2$ and $Y = Y_1 \cup Y2$, with each Y_i fibered elliptically over X_i and fibered with *rational elliptic surface* fibers over B. The intersection $Y_1 \cap Y_2$ is then the heterotic Calabi-Yau Z. In this limit the birational isomorphism of the moduli spaces can be proved geometrically [CD, D5]. This combines the description of $\mathcal{M}_Z^G$ in terms of data on B, as outlined in Section 3.8.2, with an analysis of the intermediate Jacobian and Deligne cohomology group of rational elliptic surface fibrations, extending earlier work of Kanev [K]. The picture which emerges is of an equivalence of two integrable systems: one is the version of the meromorphic Higgs bundles applicable to moduli of bundles on elliptic fibrations, the other a system of intermediate Jacobians over the moduli of Calabi-Yaus [DM2]. Some information on the extension of this birational isomorphism to various geometric strata of the moduli was obtained in [AD] using the non-linear deformation picture of [DEL], and a computation of the vanishing cycles as we degenerate from a general point of the F-theory moduli space into the geometric limit was done in [A]. Still, the conjectured isomorphism away from this geometric limit remains elusive, and even understanding the precise picture along all the strata inside the geometric limit should be very useful and pretty.

Interpretation of the trigonometric case
There are three types of connected, 1-dimensional groups: an elliptic curve E, the multiplicative group $G_m := \mathbf{C}^*$, and the additive group $G_a := \mathbf{C}$. The latter two can be considered as the groups of non-singular points in degenerations of the elliptic curve E with 1 or 2 vanishing cycles, respectively, i.e. degenerations of E to a curve with a node or a cusp. The natural functions on these three types of groups are, respectively, elliptic (doubly periodic), trigonometric (singly periodic), and rational.

Hitchin's system on the moduli of bundles on an arbitrary curve X, as well as Markman's meromorphic extension, involve Higgs bundles with arbitrary structure group G but with values in a line bundle, which is a group variety over X with fiber G_a. The moduli space of G-bundles on an elliptic fibration involves Higgs bundles with structure group G and with values in the elliptic fibration. Is there an interesting geometric interpretation of the remaining "trigonometric" case, where the values are taken in G_m?

Abelian solitons
An abelian soliton is a k-dimensional abelian subvariety $A \subset Jac(C)$ of a g-dimensional Jacobian, having the property that the kth osculating subspace to the Abel-Jacobi image $AJ(C)$ at some point $p \in C$ equals the tangent space $T_{AJ(p)}A$. This is equivalent to saying that the first k flows of Krichever's KP hierarchy solution corresponding to C, which ordinarily would evolve on the entire $Jac(C)$, happen to be confined to A. The case $k = 1$ is that of elliptic solitons, and the complementary case $k = g - 1$ of *coelliptic solitons* was studied in [DP]. In both cases,

the Jacobians or Pryms fit together to form an integrable system: a symplectic leaf of Markman's system for the elliptic solitons, and a certain non-linear reduction for the coelliptic solitons. Are there abelian solitons in other dimensions? Do they form integrable systems?

Polygonal systems

There are a number of other algebraically integrable systems which are probably related to meromorphic Higgs bundles, although this relation has not been worked out yet. One particularly beautiful example is the system of closed space polygons [Kl, KM]. An $(n + 3)$-gon in $\mathbf{R}^3$ (or, for that matter, in $\mathbf{C}^3$) with a vertex at the origin can be specified by listing an ordered $(n + 3)$-tuple of vectors $v_i, i = 1, \ldots, n + 3$ satisfying $\sum_{i=1}^{n+3} v_i = 0$. The moduli space of these polygons (modulo the action of $SO(3)$) is a $3n + 3$-dimensional Poisson variety, with $2n$-dimensional symplectic leaves obtained by fixing the lengths of the $n + 3$ sides. This is easiest to see by identifying affine 3-space with the Lie algebra $\mathbf{so}(3)$ as we did in Section 3.2.2. The configuration space of polygons is then the reduction $(\mathbf{so}(3))^{n+3}//SO(3)$, and its symplectic leaves are obtained by fixing the Casimirs of the individual $\mathbf{so}(3)$ factors, i.e. the side lengths. What is a little less obvious is that you get an algebraically integrable system by taking as Hamiltonians the lengths of the n *diagonals*

$$d_j := \sum_{i=1}^{j} v_i, \qquad j = 2, \ldots, n + 1.$$

The flow on the space of polygons corresponding to the jth Hamiltonian is simply the rotation of the first j vertices and edges around the jth diagonal. It is not known whether this system can be obtained by choosing the residues appropriately in some version of Markman's system.

References

[A]   Aspinwall: Aspects of the hypermultiplet moduli space in string duality, *JHEP* 9804 (1998) 019, hep-th/9802194.

[AD] Aspinwall, Donagi: The heterotic string, the tangent bundle, and derived categories, *Adv. Theor. Math. Phys.* 2 (1998) 1041-1074, hep-th/9806094.

[AHH] Adams, Harnad, Hurtubise: Isospectral Hamiltonian flows in finite and infinite dimensions, II: Integration of flows, *Comm. Math. Phys.* 134 (1990), 555-585.

[At] Atiyah: Vector bundles over an elliptic curve, *Proc. Lond. Math. Soc.* 7 (1957), 414-452.

[Be] Beauville: Jacobiennes des courbes spectrales et systèmes hamiltoniens complètement intégrables, *Acta Math.* 164 (1990), 211-235.

58

[Bo] Bottacin: Thesis, Orsay, 1992.

[BK] Beilinson, Kazhdan: Flat projective connections, unpublished (1990).

[CG] Clemens, Griffiths: The intermediate Jacobian of the cubic threefold, *Ann. of Math.* 95 (1972), 281-356.

[CD] Curio, Donagi: Moduli in N=1 heterotic/F-theory duality, *Nucl. Phys.* B518 (1998), 603-631. hep-th/9801057.

[D1] Donagi: Spectral covers, in: Current topics in complex algebraic geometry, MSRI 28(1992), 65-86. (alg-geom/9505009).

[D2] Donagi: Seiberg-Witten integrable systems, in: Proceedings, Santa Cruz Alg. Geom. Conf.; reprinted in: *Surveys in Differential Geometry*, ed. S.T.Yau

[D3] Donagi: Principal bundles on elliptic fibrations, *Asian J. Math.* Vol. 1 (June 1997), 214-223. alg-geom/9702002.

[D4] Donagi: Taniguchi lecture on principal bundles on elliptic fibrations, in: Integrable Systems and Algebraic Geometry; Saito, Shimizu and Ueno, eds. *World Sci.* (1998). hep-th/9802094.

[D5] Donagi: ICMP lecture on Heterotic/F-theory duality. hep-th/9802093.

[DEL] Donagi, Ein, Lazarsfeld: Nilpotent cones and sheaves on K3 surfaces. alg-geom/9504017.

[DG] Donagi, Gaitsgory: The gerbe of Higgs bundles. math. AG/0005132

[DHP] D'Hoker, Phong: Spectral curves for super-Yang-Mills with adjoint hypermultiplet for general Lie algebras, *Nucl. Phys.* B534 (1998), 697-719, hep-th/9804126.

[DM] Donagi, Markman: Spectral curves, algebraically completely integrable Hamiltonian systems, and moduli of bundles, in: Integrable Systems and Quantum Groups, LNM 1620 (1996), 1-119. alg-geom/9507017.

[DM2] Donagi, Markman: Cubics, integrable systems, and Calabi-Yau threefolds, in: Proceedings of the Algebraic Geometry Workshop on the Occasion of the 65th Birthday of F. Hirzebruch, 1993. alg-geom/9408004.

[DP] Donagi, Previato: Abelian solitons. nlin.SI/0009004.

[DW] Donagi, Witten: Supersymmetric Yang-Mills systems And integrable systems, *Nucl. Phys.* B460 (1996) 299-334. hep-th/9510101.

[F] Faltings: Stable G-bundles and projective connections, *J. Algebr. Geom.* 2 (1993), 507-568.

[FMW] Friedman, Morgan, Witten: Vector bundles and F theory, *Commun. Math. Phys.* 187 (1997) 679-743. hep-th/9701162.

[GH] Griffiths, Harris: Principles of algebraic geometry, *Pure and Applied Mathematics.* Wiley-Interscience (1978).

[GS] Gomez, Sharpe: D-branes and scheme theory. hep-th/0008150

[H] Hitchin: Stable bundles and integrable systems, *Duke* 54 (1987), 91-114.

[K] Kanev: Spectral curves, simple Lie algebras and Prym–Tjurin varieties, *Proc. Symp. Pure Math.* 49 (1989), Part I, 627-645.

[Kl] Klyachko: Spatial polygons and stable configurations of points on the projective line, *Alg. Geom. and its Applns,* Yaroslavl (1992), 67-84.

[KM] Kapovich, Millson: The symplectic geometry of polygons in Euclidean space, *J. Differ. Geom.* 44 (1996), 479-513.

[Kr] Krichever: Elliptic solutions of the Kadomtsev–Petviashvili equation and integrable systems of particles, *Functional Anal. Appl.* 14 (1980), 282-290.

[La] Laumon: Un analogue global du cône nilpotent, *Duke* 57 (1988), 647-671.

[Lo] Looijenga: Root systems and elliptic curves, *Inv. Math.* 38 (1976), 17-32.

[M] Markman: Spectral curves and integrable systems, *Comp. Math.* 93 (1994), 255-290.

[Mu] Mukai: Symplectic structure of the moduli space of sheaves on an abelian or K3 surface, *Inv. Math.* 77 (1984), 101-116.

[Mum] Mumford: Tata Lectures on Theta II, Birkhaeuser-Verlag, Basel, Switzerland and Cambridge, MA (1984).

[MV] Morrison, Vafa: Compactifications of F-theory on Calabi–Yau threefolds, I: *Nucl. Phys.* B473 (1996) 74-92. hep-th/9602114; and II: *Nucl. Phys.* B476 (1996) 437-469. hep-th/9603161.

[S] Scognamillo: An elementary approach to the abelianization of the Hitchin system for arbitrary reductive groups. alg-geom/9412020.

[SW1] Seiberg, Witten: Monopole condensation, and confinement In N=2 supersymmetric Yang-Mills theory, *Nucl. Phys.* B426 (1994) 19-52; Erratum-ibid. B430 (1994) 485-486. hep-th/9407087.

[SW2] Seiberg, Witten: Monopoles, duality and Chiral symmetry breaking in N=2 supersymmetric QCD, *Nucl. Phys.* B431 (1994), 484-550. hep-th/9408099.

[T] Tyurin: Symplectic structures on the varieties of moduli of vector bundles on algebraic surfaces with $p_g > 0$, *Math. USSR Izvestiya* 33 (1989).

[TV] Treibich, Verdier: Solitons elliptiques. *The Grothendieck Festschrift*, Vol. III, 437-480, Birkhauser Boston (1990).

[V] Vafa: Evidence for F-Theory, *Nucl. Phys.* B469 (1996), 403-418. hep-th/9602022.

4

The anti-self-dual Yang–Mills equations and their reductions

Lionel Mason

The Mathematical Institute, Oxford
`lmason@maths.ox.ac.uk`

Abstract

These notes provide an introduction to an approach to the theory of integrable systems that arises from the observation that many integrable systems are reductions of the anti-self-duality equations so that the theory of these equations can be understood as a reduction of the corresponding theory for the anti-self-duality equations.

We start with a general discussion of integrable systems and relations between them arising from symmetry reductions and give some standard examples. We then give an introduction to gauge theory and the self-dual Yang–Mills equations.

The anti-self-dual Yang–Mills equations can be seen to be an integrable system; it has a Lax pair, admits Backlund transformations, there are ansatze for solutions, and it has topological solutions, instantons. We go on to discuss its reductions to three dimensions, monopoles and Chiral models.

Another way to see the self-dual Yang–Mills equations as an integrable system is to present Hamiltonian and Lagrangian formulations and a recursion operator. This leads to the generalised self-dual Yang–Mills Hierarchies.

We then discuss general principals of reduction. Translation reductions lead to the KdV and nonlinear Schrodinger equations. Non translational reductions give rise to the Ernst equations and Painlevé equations. Finally we discuss further developments of the overview.

4.1 Introduction

These notes are intended to provide an introduction to an overview on the theory of integrable systems based on reductions of the anti-self-dual Yang–Mills equations and its twistor construction. This overview was originally proposed by Ward (1985) and developed in Mason & Woodhouse (1996) and this latter book contains full details of most of the material presented here and much more. These notes are intended to be accessible to graduate students and so introductory material on connections and the Yang–Mills equations is provided, together with some exercises at the end.

4.1.1 Standard aspects of integrability

Integrable systems are differential equations that, despite their nontrivial nonlinearity, are surprisingly tractable. The following more precise definitions do not always apply:

- **Theorem 1 (Arnol'd-Liouville)** *Suppose that a Hamiltonian system in 2n-dimensions has n constants of the motion H_i in involution such that the map to $\mathbb{R}^n$ determined by the H_i is proper and regular, then there exists a coordinate system of 'action-angle' variables, obtainable by quadratures, in which the flows are linear.*

- **Theorem 2 (Magri)** *If a Hamiltonian system admits a recursion operator satisfying certain conditions, then the system is integrable in the Arnol'd-Liouville sense.*
- The system should admit a twistor correspondence.
- The system is the consistency condition for a 'Lax pair', i.e. an auxilliary system of overdetermined linear equations.

4.1.2 Properties of solutions to integrable systems

Integrability usually means that explicit formulae for solutions are readily obtainable. Such solutions have characteristic properties

- **The Painlevé property:** for ODE's, solutions to linear equations are always regular (except at the fixed singularities of the equations). For complex nonlinear equations, solutions generically have branching and essential singularities or worse. The Painlevé property requires that all *moveable* singularities be rational, i.e. any singularity whose location depends on the initial conditions must be rational.

It is conjectured, but not yet proved, that is a defining property.

- **Solitons:** integrable equations often admit 'particle-like' lump solutions, solitons. These can be superposed and the lumps often simply pass through each other, perhaps with some simple scattering, but maintaining their integrity.

- **IST:** the inverse scattering transform expresses the general solution as a nonlinear superposition of radiative or dispersive modes, and solitons.

4.1.3 Key examples

Because of the imprecise nature of the definition of an integrable system, it is important to familiarize oneself with the key examples in order to get a feel for the subject.

The Euler top

The configuration space is $SO(3)$ ={frames fixed in the top}, and the phase space is $T^*SO(3)$. The moving frame $\mathbf{e}_i$, $i = 1, 2, 3$ satisfies

$$\frac{\mathrm{d}\mathbf{e}_i}{\mathrm{d}t} = \boldsymbol{\omega} \wedge \mathbf{e}_i, \qquad \boldsymbol{\omega} = \omega_1 \mathbf{e}_1 + \omega_2 \mathbf{e}_2 + \omega_3 \mathbf{e}_3,$$

and, for the Euler top with principal moments of inertia I_i (constants)

$$I_1 \frac{\mathrm{d}\omega_1}{\mathrm{d}t} = (I_2 - I_3)\omega_2\omega_3, \qquad + \qquad \text{cyclic.}$$

Constants of motion: the motion is generated by the Hamiltonian $\frac{1}{2}\sum_i I_i\omega_i^2$ and there are two other constants of the motion in involution; the total angular momentum $\mathbf{L} \cdot \mathbf{L} = \sum_i I_i^2\omega_i^2$ and $\mathbf{L} \cdot \mathbf{k}$, $\mathbf{L} = \sum_i I_i\omega_i\mathbf{e}_i$.

The equations for ω_i can be rescaled to give

$$\dot{\omega}_1 = 2\omega_2\omega_3, \qquad \dot{\omega}_2 = -2\omega_3\omega_1, \qquad \dot{\omega}_3 = 2\omega_1\omega_2.$$

Lax pair: these equations are equivalent to the integrability, $[L_0, L_1] = 0$ of the Lax pair

$$L_0 = \frac{\mathrm{d}}{\mathrm{d}t} - i(\Omega_3 + \lambda(\Omega_1 + i\Omega_2))$$

$$L_1 = \lambda(\Omega_1 + i\Omega_2) + 2\Omega_3 - (\Omega_1 - i\Omega_2)/\lambda$$

where

$$\Omega_1 = \begin{pmatrix} 0 & \omega_1 \\ \omega_1 & 0 \end{pmatrix}, \ \Omega_2 = \begin{pmatrix} 0 & -\omega_2 \\ \omega_2 & 0 \end{pmatrix}, \ \Omega_3 = \begin{pmatrix} \omega_3 & 0 \\ 0 & \omega_3 \end{pmatrix}.$$

Note that L_1 evolves by conjugation, so that $\text{tr}(L_1)^p$ is constant. This yields the conserved quantities.

The Korteweg–de Vries equations

This is an equations for the evolution of the height $u(x,t)$ of shallow water waves in a channel

$$u = u(x,t), \qquad 4u_t - u_{xxx} + 6uu_x = 0.$$

This equation is equivalent to the integrability conditions $[L_0, L_1] = 0$ for the Lax pair

$$L_0 = \partial_x + \begin{pmatrix} q & -1 \\ p & -q \end{pmatrix} - \lambda \begin{pmatrix} 0 & 0 \\ 1 & 0 \end{pmatrix}, \qquad L_1 = \partial_t + B - \lambda \partial_x$$

where p and B are determined in terms of q by the consistency conditions and $u = -2q_x$.

Solitons: this equation admits 'soliton' solutions, in particular the one soliton is

$$u = 2c \cosh^{-2}(c(x - ct)), \qquad c = \text{ velocity}.$$

There exist n-soliton solutions with n lumps moving at different speeds that keep their identity under interaction.

The Kadomtsev–Petviashvilii equations

These equations control the evolution of shallow water waves in 2-dimensions

$$u := u(x,y,t), \qquad \partial_x(4u_t - u_{xxx} + 6uu_x) = u_{yy}.$$

The Lax pair is

$$L_0 = \partial_y - \partial_x^2 - 2u, \qquad L_1 = \partial_t - \partial_x^3 + 3u\partial_x + v$$

where the consistency conditions determine's v in terms of u in addition to forcing the KP equation.

4.1.4 Reductions of integrable systems

Reduction can mean the imposition of a symmetry or the specialization of a parameter in a system of equations. For example imposing a symmetry along ∂_y reduces the KP equation to the KdV equation.

Generally, reduction is compatible with the structures such as the Lax Pair associated with integrability and so yield an integrable system. Hence it gives a partial ordering on the set of integrable equations.

Is there a universal integrable system? The answer is almost certainly no; one can construct integrable systems in arbitrarily high dimensions for example. However, the anti-self-dual Yang–Mills (ASDYM) equations reduce to many of the most popular integrable systems. Its flexibility lies in the fact that it is really a family of equations, one for each choice of gauge group.

This leads to two programmes:

(i) classify those integrable systems that can be obtained by reduction from the ASDYM equations;

(ii) unify the theory of these equations by reduction of the corresponding theory for the ASDYM equations.

This leads to a self contained theory restricted to these systems, but also highlights the distinctions from systems that are not reductions of ASDYM. In particular the KP equations do not appear to be a reduction of ASDYM with finite dimensional gauge group, although in Ablowitz & Clarkson (1991) it is shown to be a reduction if an infinite-dimensional gauge group is allowed. (This may not, however, mean so much as it is not so clear that the ASDYM equations are integrable in such a meaningful way with an infinite-dimensional gauge group.)

4.1.5 Twistor theory

The basic aim of Penrose's twistor theory is to find 1–1 correspondences between

$$\left\{ \begin{array}{l} \text{Solutions to physical equations; Yang–Mills, Einstein} \end{array} \right\} \longleftrightarrow \left\{ \begin{array}{l} \text{Deformed complex structures on twistor space} \end{array} \right\}$$

The hope has been that twistor space provides the correct geometric arena for the correct formulation of quantum gravity. Evidence for the existence of such constructions comes from the twistor correspondences for the self-duality equations and these suggest the larger (unfulfilled) programme for the full Yang–Mills and Einstein equations.

These twistor correspondences for the self-duality equations provide a paradigm for 'complete integrability'. They give a geometric construction for the general local solution to the equations. The fact that the self-duality equations yield 'most' integrable systems under symmetry reduction implies that these reductions also inherit reduced twistor correspondences. This suggests that it is the existence of a twistor construction that underlies integrability. This can be more general than merely looking at reductions of self-dual Yang–Mills as many more equations, in arbitrarily high dimension for example, admit a twistor correspondence.

These notes are too brief to give an introduction to Twistor theory and the interested reader is referred to Mason & Woodhouse (1996) part II for a full treatment of these ideas. However, it is the twistor theory that is the real impetus for many of these ideas.

4.2 The self-dual Yang–Mills equations

We need some background geometry first. The Yang–Mills equations depend on the choice of a Lie group, G. When $G = U(1)$ they reduce to Maxwell's equations. The anti-self-duality condition picks out circularly polarized solutions. The full equations make good sense for space-times of any signature and dimension. However, the self-duality condition only makes sense in dimension four, and only admits real solutions when the signature is Euclidean or ultrahyperbolic, $(+ + --)$.

4.2.1 Bundles, connections and curvature

The Yang–Mills equations are equations on connections on vector-bundles. A vector bundle on a manifold M is an attachment of a vector space E_x to each $x \in M$. More formally:

Definition 1 *A vector bundle of rank n over a manifold M is a manifold E fibred over M, $p : E \to M$ such that the fibres $p^{-1}(x)$, $x \in M$ have the structure of n-dimensional real or complex vector spaces depending smoothly on $x \in M$ (sometimes holomorphically if the fibres and M are complex in which case the bundle will be said to be holomorphic).*

A complex vector bundle is unitary if each fibre has a Hermitean metric (depending smoothly on $x \in M$) and special unitary if there is a holomorphic volume form on each fibre.

A bundle is usually described by means of *local trivializations* associated to a cover by topologically trivial open sets $U_i \subset M$, i.e. isomorphisms $\rho_i : E|_{U_i} \cong U_i \times \mathbb{C}^n$. On overlaps, $U_1 \cap U_2$, there will be an $n \times n$ matrix function $\rho_{12} = \rho_1 \rho_2^{-1}$ satisfying, for consistency $\rho_{12} \rho_{23} = \rho_{13}$ on triple overlaps. A vector bundle is said to have structure group G if the ρ_{ij} can be chosen so as to take values in G. One can take direct sums, $E_1 \oplus E_2$, duals E^* and tensor products $E_1 \otimes E_2$ of vector bundles fibrewise over each $x \in M$ following the definitions for vector spaces. The bundle $E \otimes E^*$ is of particular note as the bundle $\mathrm{End}(E)$ of endomorphisms of E.

Example. A simple nontrivial example is TS^2, the tangent bundle of the sphere; $T_x S^2$ is the space of tangent vectors at $x \in S^2$. This is also a holomorphic complex line bundle and the structure group can be reduced to $SO(2) = U(1)$ (exercise).

A gauge transformation g is an automorphism, i.e. a diffeomorphism $g : E \to E$ that sends each fibre to itself by a linear transformation (in G if the structure group is G). In a local trivialization it will be represented by a matrix function $g(x)$, $x \in M$ on U_i with values in G. Even when the bundle is topologically trivial, $E = M \times \mathbb{C}^n$, the bundle concept is nontrivial in the sense that the given trivialization is not part of the structure and any gauge transformation maps this to a different but equivalent trivialization.

A section $s \in \Gamma(E, M)$ is a smooth map $s : M \to E$. These can be added together and multiplied by functions on M. We would like a definition of differentiation of sections that does not depend on a choice of (local) trivialization.

Definition 2 *A connection on a bundle E is a linear map D from sections of E to sections of $T^* M \otimes E$ such that $D(fs) = (\mathrm{d}f)s + fDs$ where f is a function on M and d is the exterior derivative.*

In a local trivialization ρ_i, put $v = \rho_i(s)$, here v is an n-component column vector, then

$$\rho_i(Ds) = \mathrm{d}x^a \left(\frac{\partial}{\partial x^a} + A_a^i \right) \rho_i(s) = (d + A)v, \quad A = \mathrm{d}x^a A_a$$

where x^a are coordinates on M and $A_a^i = \rho_i D_a \rho_i^{-1}$ are matrices. Usually one misses out the ρ_i in the above formula working instead with a function v with values in $\mathbb{C}^n$ and writes $Dv = \mathrm{d}v + Av$. Under a change of local trivialization, we must have

$$A_a^i = \rho_{ij} A_a^j \rho_{ij}^{-1} + \rho_{ij} \mathrm{d}\rho_{ij}^{-1}\,.$$

Gauge transformations send D to $g \circ D \circ g^{-1} = D + (gDg^{-1})$, the same formula as above if expressed in a frame. When the bundle has structure group G, then the A_a are usually taken to take values in $Lie(G)$, the Lie algebra T_eG of G.

The connection naturally extends to E^* and $E \otimes E^*$ etc by requiring the Leibnitz (product) rule over contractions and tensor products. For a section γ of $E \otimes E^*$ we will have, in a local trivialization $\rho_i(D\gamma) = d\rho_i(\gamma) + [A, \rho_i(\gamma)]$.

The curvature is a 2-form with values in $E \otimes E^*$ defined to be $F = D \wedge D$ or in indices in a local trivialization

$$F_{ab} = \partial_{[a} A_{b]} + [A_a, A_b]\,.$$

The curvature transforms homogeneously under gauge transformations (and changes of trivialization) $F \to g^{-1}Fg$ and satisfies the Bianchi identity

$$D \wedge F = 0\,, \qquad \text{or in indices} \qquad D_{[a} F_{bc]} = 0$$

which follows from the Jacobi identity $[D_a, [D_b, D_c]] = 0$.

4.2.2 The full Yang–Mills equations

The Yang–Mills equations arise from the Lagrangian density $L = -\mathrm{tr}(F_{ab} F^{ab})$ where M is now assumed to be endowed with a metric which is used to raise and lower indices and tr is shorthand for a negative definite invariant inner product on $Lie(G)$ (which is indeed the trace for $SU(n)$). The Euler-Lagrange equations are $D^a F_{ab} = 0$.

When $G = U(1)$ and M is Minkowski space, this gives Maxwell's equations by identifying $F_{0j} = iE_j$ the electric field, and $F_{jk} = i\varepsilon_{jkl}B_l$ the magnetic field, $j, k, l = 1, 2, 3$.

Note that the equations are gauge invariant so that if D is a solution to the Yang–Mills equations, then so is $g \circ D \circ g^{-1}$. This connection is regarded as equivalent to D and one is in general interested only in gauge equivalence classes of solutions.

Unlike Maxwell's equations, the Yang–Mills equations only acquire full relevance for physics as a quantum field theory in which they describe the weak and the strong nuclear interactions. However, they make perfectly good sense as a set of differential equations. Nevertheless this fact means that solutions to the equations in Euclidean signature have more physical significance than one might expect on account of the Euclidean path integral approach to quantum field theory. In particular, solutions that are absolute minima of the action in Euclidean signature play a significant role as 'instantons', vacua about which the theory can be expanded.

4.2.3 Anti-self-duality and the different signatures

Minkowski space $\mathbb{M}$ is $\mathbb{R}^4$ together with a metric $\eta_{ab} = \mathrm{diag}(1, -1, -1, -1)$ and volume form $\varepsilon_{abcd} = \varepsilon_{[abcd]}$, $\varepsilon_{0123} = 1$. Euclidean space $\mathbb{E}$ and ultrahyperbolic space $\mathbb{U}$ have the volume form and metrics of signature $(+ + + +)$ and $(+ + - -)$ respectively.

On 2-forms, F_{ab} we can define the duality operation

$$F_{ab}^* = \frac{1}{2}\varepsilon_{abcd}F^{cd}$$

and it can be seen (exercise) that $F^{**} = -F$ in Minkowski signature, whereas $F^{**} = F$ in Euclidean and ultra hyperbolic signature. Thus the eigenvalues are $\pm i$ in Minkowski signature, but ± 1 in Euclidean and ultrahyperbolic signature. The duality operation is also conformally invariant on 2-forms. In terms of two component spinors (cf. Huggett & Tod 1994) the decomposition into eigenspaces of $*$ corresponds to the spinor decomposition of 2-forms

$$F_{AA'BB'} = \varepsilon_{AB}\phi_{A'B'} + \varepsilon_{A'B'}\phi_{AB},$$
$$2\phi_{AB} = F_{AA'B}{}^{B'}, \quad 2\phi_{A'B'} = -F^A{}_{A'BB'},$$

where the first term is self-dual and the second anti-self-dual. We often write $F = F^+ + F^-$ for the decomposition of a 2-form into its SD and ASD parts.

The anti-self-dual Yang–Mills equations are the conditions that $F^* = -F$ in Euclidean or ultra-hyperbolic signature, or $F^* = -iF$ in Minkowski signature, i.e. $F = F^-$.

Solutions to the ASDYM equations are also solutions to the full YM equations as the full YM equations can be written as $D \wedge F^* = 0$, but $D \wedge F^* = -D \wedge F$ by anti-self-duality and $D \wedge F = 0$ by the Bianchi identity. Note also that the equations are conformally invariant (like the full Yang–Mills equations) as the $*$ operator is conformally invariant on 2-forms in four dimensions.

If one changes the sign of the volume form, a self-dual 2-form becomes anti-self-dual and vice-versa. Thus there is not much to choose between the self-dual and anti-self-dual Yang–Mills equations. However, when one has also made a choice of complex structure on $\mathbb{E}$ or $\mathbb{U}$, there is a natural choice of orientation and the anti-self-dual Yang–Mills connections are automatically also holomorphic. For this reason it is more usual to work with the anti-self-dual Yang Mills equations.

4.2.4 Anti-instantons

Definition 3 *Instantons are defined to be the absoluted minima of the action* $S = \int_{\mathbb{R}^4}(F,F)\nu$ *where* $(F,F) = -\mathrm{tr}(F_{ab}F^{ab})$ *is a positive definite metric on the curvature and the signature is taken to be Euclidean.*

These relate to the ASDYM equations on S^4 since, firstly a theorem of Uhlenbeck's shows that if the action is finite, then the gauge field extends to $S^4 = \mathbb{R}^4 \cup \{\text{point}\}$ (the identification here is given by stereographic projection), Uhlenbeck (1982). Further the action

$$S = \int_{S^4}(F,F)\nu = \int_{S^4}((F^+,F^+)+(F^-,F^-))\nu\,,$$

whereas

$$8\pi^2 k = \int_{S^4}((F^+,F^+)-(F^-,F^-))\nu$$

is a topological invariant, the second Chern class, or instanton number. Assuming this to be negative (we wish to work with ASD fields)

$$S = -8\pi^2 k + 2\int_{S^4}(F^+,F^+)\nu$$

so that $F^+ = 0$ clearly gives absolute minima for a bundle in a given topological class (with k negative).

Any solution of the ASDYM equations on $\mathbb{R}^4$ with finite action is therefore an anti-instanton. For a full discussion of instantons and their twistor theory, see Atiyah (1979).

Reduction by 1 symmetry

The simplest way to impose a symmetry is to drop the dependence on $x^0 = t$. In this case one must also restrict the gauge transformations so that they too do not depend on t and this implies that the component $\Phi = A^0$ transforms homogeneously under gauge transformations $\Phi \to g\Phi g^{-1}$. Then the data reduces to (D_i, Φ), where D_i is a connection on the bundle E over $\mathbb{R}^3$ and Φ is a section of $E \otimes E^*$. The ASDYM equations reduce to

$$F_{ij} = \varepsilon_{ij}{}^k D_k \Phi, \quad \text{or} \quad F =^* D\Phi$$

where now $*$ is the 3-dimensional $*$-operator. This makes sense both for a reduction from $\mathbb{E}$ and from $\mathbb{U}$, and we get equations on $\mathbb{E}^3$ and $\mathbb{R}^{2+1}$ respectively.

The natural boundary condition on $\mathbb{E}^3$ is that the 'energy'

$$E = \int_{\mathbb{E}^3} (F_{ij}, F^{ij}) + (D_i \Phi, D^i \Phi) \mathrm{d}vol_3$$

should be finite. The solutions will be trivial unless one also requires that $-\mathrm{tr}(\Phi^2) \to 1$ as $|x| \to \infty$. This characterizes the monopole solutions. These also admit topological invariants of a more subtle kind; consider the $n = 2$ case. Then the eigenspaces of Φ for large enough $|x|$ are complex line bundles on S^2 and these have integral Chern class

$$c_1 = \lim_{r \to \infty} \frac{1}{4\pi} \int_{|x|=r} \mathrm{tr}(\Phi F),$$

the monopole number.

On $\mathbb{R}^{2+1}$, the equations are evolution equations and the finite energy condition is a condition that one would like to impose on data on an initial 2-dimensional hypersurface. There are lump solutions here, but the lump number is not understood as a topological invariant.

4.3 ASDYM as an integrable system

One feature of integrable systems is the ease, despite their nonlinearity, with which one can write down exact solutions. The simplest of these for the ASDYM equations is the t'Hooft ansatz. This has a neat geometric interpretation in terms of spinors in curved space-time. Consider the curved metric $\mathrm{d}s^2 = \phi^2 \delta$ where δ is the flat Euclidean metric. The connection on the spin bundle for $\mathrm{d}s^2$ can be represented in terms of the flat metric by

$$\nabla_{AA'} \alpha_{B'} = \partial_{AA'} \alpha_{B'} - \alpha_{A'} \partial_{AB'} \log \phi + \frac{1}{2} \alpha_{B'} \partial_{AA'} \log \phi$$

where the flat space spinor $\alpha_{A'}$ has been identified with the spinor $\phi^{1/2} \alpha_{A'}$ for $\mathrm{d}s^2$. The gauge potential can be alternatively written as

$$A_{AA'}{}^{B'C'} = -\varepsilon_{A'}{}^{(C'} \partial_A{}^{B')} \log \phi,$$

or

$$A = i\boldsymbol{\sigma} \cdot \boldsymbol{\nabla} \log \phi \mathrm{d}t + i(\boldsymbol{\sigma} \wedge \boldsymbol{\nabla} \log \phi + \boldsymbol{\sigma}\partial_t \log \phi) \cdot \mathbf{dx},$$

where $\boldsymbol{\sigma} = (\sigma_1.\sigma_2, \sigma_3)$ are the Pauli matrices.

The Weyl curvature for $\nabla_{AA'}$ is zero as it is conformally flat, so the curvature is

$$[\nabla_a \nabla_b]\alpha_{C'} = -\varepsilon_{A'B'}\Phi_{ABC'}{}^{D'}\alpha_{D'} - 2\Lambda\varepsilon_{AB}\alpha_{(A'}\varepsilon_{B')C'}$$

and one can see that if $4\Lambda = \partial^a\partial_a \log \phi + (\partial^a \log \phi)(\partial_a \log \phi) = 0$ the connection is ASD. However, this will be the case if $\Box\phi = 0$. Thus, any non vanishing solution of the Laplacian on $\mathbb{E}$ gives rise to a solution of the ASDYM equation (locally) so long as $\phi \neq 0$.

The solutions $\phi = 1 + \sum_{i=1}^{k} \lambda_i/|x - a_i|^2$, $\lambda_i > 0$ are particularly significant as the singularities in ϕ do not lead to singularities in the ASDYM field and they are nowhere vanishing and lead to anti-instanton solutions on S^4 with instanton number k as described above. These are not all the instanton solutions except for $k = 1, 2$. In general there is an $8k - 3$-dimensional moduli space but the other solutions are harder to describe, although they do have a complete description in terms of some nonlinear algebraic equations via the 'Atiyah–Drinfeld–Hitchin–Manin (ADHM)' construction, see Atiyah (1979).

4.3.1 The ASDYM Lax pair

The spinor correspondence leads to coordinates in which the anti-self-duality condition can be seen to be a flatness condition on a certain family of 2-planes called α-planes.

$$x^{AA'} = \frac{1}{\sqrt{2}}\begin{pmatrix} x^0 + x^1 & x^2 - ix^3 \\ x^2 + ix^3 & x^0 - x^1 \end{pmatrix} = \frac{1}{\sqrt{2}}(tI + \mathbf{r} \cdot \boldsymbol{\sigma}) = \begin{pmatrix} \tilde{z} & w \\ \tilde{w} & z \end{pmatrix}$$

where the x^a are rectilinear Minkowski space coordinates. In the $(z, w, \tilde{z}, \tilde{w})$ coordinates, the metric and volume form are

$$ds^2 = 2(\mathrm{d}z\mathrm{d}\tilde{z} - \mathrm{d}w\mathrm{d}\tilde{w}), \quad \nu = \mathrm{d}w \wedge \mathrm{d}\tilde{w} \wedge \mathrm{d}z \wedge \mathrm{d}\tilde{z}.$$

We will usually consider either Euclidean or ultra hyperbolic signature which are obtained with $\tilde{z} = \bar{z}$ and $\tilde{w} = \mp\bar{w}$ respectively. With these reality conditions this choice of coordinates expresses the ds^2 as a (pseudo-) Kähler metric with (pseudo-)Kähler form

$$i\omega = \mathrm{d}w \wedge \mathrm{d}\tilde{w} - \mathrm{d}z \wedge \mathrm{d}\tilde{z}$$

where the reality conditions $\tilde{z} = \bar{z}$, $\tilde{w} = \mp\bar{w}$ are imposed for $\mathbb{E}$, $\mathbb{U}$ respectively. Ultrahyperbolic signature is also obtained with $(z, w, \tilde{z}, \tilde{w})$ all real.

Define the vector fields l and m by

$$l = \partial_w - \lambda\partial_{\tilde{z}}\,, \quad m = \partial_z - \lambda\partial_{\tilde{w}},.$$

Then the bivector $l \wedge m$ is SD for all λ and varies over all such as λ varies over the Riemann sphere. Such two-planes are called α-planes. These are generally complex 2-planes, except in $\mathbb{U}$ where they can be real when $(z, w, \tilde{z}, \tilde{w})$ and λ are all real.

Self-dual bivectors are orthogonal to ASD 2-forms and indeed the ASD 2-forms F are characterized by the condition that $F(l, m) = 0$. This gives:

Proposition 1 *The ASD Yang–Mills equations on a connection are equivalent to the condition that it is flat on α-planes, i.e. that*

$$[D_w - \lambda D_{\tilde{z}}\,, D_z - \lambda D_{\tilde{w}}] = F(l, m) = 0.$$

Note that this integrability condition is equivalent to the existence of solutions to the equations

$$(D_w - \lambda D_{\tilde{z}})\psi = (D_z - \lambda D_{\tilde{w}})\psi = 0.$$

where $\psi = \psi(x^a, \lambda)$ is a basis of E at each point.

4.3.2 Backlund transformations

One of the first applications of a Lax pair is to perform Backlund transformations. One approach to these is to perform a gauge transformation to the Lax pair,

$$(D_w - \lambda D_{\tilde{z}}\,, D_z - \lambda D_{\tilde{w}}) \to (D'_w - \lambda D'_{\tilde{z}}\,, D'_z - \lambda D'_{\tilde{w}})$$
$$= g \circ (D_w - \lambda D_{\tilde{z}}\,, D_z - \lambda D_{\tilde{w}}) \circ g^{-1}$$

where g depends on λ. This is well defined and will lead to a new solution to the ASDYM equations if $g \circ (D_w - \lambda D_{\tilde{z}}\,, D_z - \lambda D_{\tilde{w}}) \circ g^{-1}$ is linear in λ; it must necessarily still commute and so will determine an ASDYM solution since, if a Lax pair with its linear dependence on λ is given, the connection coefficients can be reconstructed from it.

An example is

$$g = \begin{pmatrix} \frac{1}{\lambda - \mu} A & 0 \\ 0 & \tilde{A} \end{pmatrix}$$

for some 2×2 block decomposition relative to $k + \tilde{k} = n$, A being $k \times k$ and $\tilde{A}$ being $\tilde{k} \times \tilde{k}$ and μ being a complex constant. In general it is necessary to solve linear differential equations in order to find a gauge and A for which this works. These are already solved if we are given a solution $\psi(x^a, \lambda)$ to the Lax pair.

We illustrate the technique in the case $k = \tilde{k} = 1$, $n = 2$, and $g = \mathrm{diag}(1, \lambda)$. If we put

$$\Phi = \begin{pmatrix} a & b \\ c & d \end{pmatrix}, \quad \text{then} \quad g\Phi g^{-1} = \begin{pmatrix} a & \lambda^{-1}b \\ \lambda c & d \end{pmatrix}.$$

We need to ensure that there are no terms of order λ^{-1} or λ^2 in the transformed Lax pair. This will be the case if we can find a gauge in which A_w and A_z are lower triangular and $A_{\tilde{w}}$ and $A_{\tilde{z}}$ are upper triangular. We can do this by performing a gauge transformation to a new frame (e_1, e_2) such that $D_z e_1 \propto D_w e_1 \propto e_1$ and $D_{\tilde{z}} e_2 \propto D_{\tilde{w}} e_2 \propto e_2$. The existence of such e_1 and e_2 follows from the integrability conditions $[D_z, D_w] = 0 = [D_{\tilde{z}}, D_{\tilde{w}}]$, and indeed, if we have a solution ψ to the Lax pair that is regular both near $\lambda = 0$ and $\lambda = \infty$, then we can take one column at $\lambda = 0$ for e_1 and the other column for e_2 near $\lambda = \infty$. In this gauge one can now perform the 'singular gauge transformation' and obtain a new solution. Moreover, one can continue the process since, in the new gauge, ψ is lower triangular by construction at $\lambda = 0$ and upper triangular at $\lambda = \infty$ so that $\psi' = g\psi g^{-1}$ is regular at $\lambda = 0, \infty$ and satisfies the Lax pair with the new potentials. Thus one has the necessary ingredients to perform the procedure again and again.

These Backlund transformations are often thought of as 'hidden symmetries', they are transformations from one solution to another that do not simply arise from point transformations of space-time. However, one can understand them as an (infinite dimensional) group of symmetries and investigate their linearized action.

4.3.3 Solution generation

The above works even if we start with the trivial solution and one can, instead of iterating, work harder to construct a more general g. This can be presented somewhat differently as follows. Note that if one is given ψ, then the connection is determined by

$$\begin{aligned} ((\partial_w - \lambda\partial_{\tilde{z}})\psi)\psi^{-1} &= A_w - \lambda A_{\tilde{z}} \\ ((\partial_z - \lambda\partial_{\tilde{w}})\psi)\psi^{-1} &= A_z - \lambda A_{\tilde{w}}. \end{aligned} \tag{4.1}$$

One approach to constructing exact solutions is to make an ansatz for ψ so that the RHS of the equations above are linear in λ thus allowing one to construct a connection A for which the Lax pair commutes. A fruitful ansatze is

$$\psi = I + \sum_{i=1}^{k} \frac{M_i(x)}{\lambda - \mu_i}.$$

If we wish the connection to be unitary, we require that $\psi(x, \lambda)\psi^*$ $(x, \mp 1/\bar{\lambda}) = \alpha I$ in $\mathbb{E}$ or $\mathbb{U}$ respectively where α is a scalar independent of λ (when the coordinates are all real on $\mathbb{U}$ one can also use $\psi(x, \lambda)\psi^*(x, \bar{\lambda}) = \alpha 1$). The condition that the RHS of equation (4.1) has no poles gives further conditions: the M_i should have rank 1, $M_i = u_i \otimes v_i$ with the u_i determined in terms of v_i (or vice versa) and the v_i depend on the coordinates only through $(\tilde{z} + \mu_i w, \tilde{w} + \mu_i z)$.

4.3.4 Recursion operators

The Arnol'd-Liouville definition of integrability can be obtained when a system has the additional structure of a *recursion operator*.

Definition 4 *Let (M^{2n}, Ω) be a finite dimensional phase space with Hamiltonian H and corresponding Hamiltonian vector field X, $X \lrcorner \Omega + dH = 0$ and Poisson bracket $\{h_1, h_2\} = \Pi^{ab}(\partial_a h_1)(\partial_b h_2)$, $\Pi = \Omega^{-1}$. A recursion operator on M is a section $R_b^a \in \Gamma(\text{End}(TM))$ such that*
(1) $a\{,\} + bR\{,\}$, where $R\{h_1, h_2\} = \Pi^{ab}R_a^c(\partial_b h_1, \partial_c h_2)$, is a Poisson structure $\forall$ constants a and b, and
(2) X is also Hamiltonian with respect to $R\{,\}$.

Proposition 2 *If a (simply connected) Hamiltonian system admits a suitably non-degenerate recursion operator, then the system is Arnol'd-Liouville integrable.*

Proof: note first that since $\{,\}$ is non-degenerate then so is $\{,\} + tR\{,\}$ for small t and corresponds to the symplectic form

$$\Omega_t = (1 - tR)^{-1}\Omega = \sum_{i=0}^{\infty} t^i R^i \Omega := \sum_{i=0}^{\infty} t^i \Omega_i$$

for all t. Thus each $\Omega_i = R^i \Omega$ is closed and non-degenerate and is therefore a symplectic form. Since X is Hamiltonian wrt $\{,\}$ and $R\{,\}$, $\mathcal{L}_X \Pi = \mathcal{L}_X R\Pi = 0$ so that $\mathcal{L}_X R = 0$ and hence $\mathcal{L}_X \Omega_i = 0$. Hence

X is Hamiltonian with respect to each Ω_i with Hamiltonian h_i, $\mathrm{d}h_i = -\Omega_i(X,) = -\Omega(R^i X,)$. Furthermore, $\{h_i, h_j\} = 0$ since

$$\{h_i, h_j\} = \Omega(R^i X, R^j X) = \Omega_{j+k}(X, X) = 0$$

by skew symmetry of Ω and the property $R^a_c \Omega_{ab} = -R^a_b \Omega_{ac}$ that follows from skew symmetry of $R\Omega$ etc..

Thus, if n of the h_i are independent, then we have an Arnol'd-Liouville integrable system (and this is what is meant by the 'suitable' in the assumption of non-degeneracy of R). □

Example. In the case of the KdV equations, $M = \mathcal{S}^\infty(\mathbb{R})$ the space of smooth functions on the line with rapid decrease at ∞.

We have two Poisson structures

$$\{F[u], G[u]\}_1 = \int_{\mathbb{R}} \frac{\delta F}{\delta u(x)} \partial_x \frac{\delta G}{\delta u(x)} \mathrm{d}x$$

$$\{F[u], G[u]\}_2 = \int_{\mathbb{R}} \frac{\delta F}{\delta u(x)} (\frac{1}{4}\partial_x^3 + u\partial_x + \frac{1}{2}u_x) \frac{\delta G}{\delta u(x)} \mathrm{d}x .$$

The Hamiltonian $h_0 = \int \frac{1}{2}u^2 \mathrm{d}x$ generates translations in x wrt $\{,\}_1$ and the KdV flow wrt $\{,\}_2$. Thus the system is bi-Hamiltonian with recursion operator

$$R = (\frac{1}{4}\partial_x^3 + u\partial_x + \frac{1}{2}u_x) \circ \partial_x^{-1}.$$

Proposition 3 (Magri, 1978) *R is a recursion operator for the KdV flow.*

We obtain higher Hamiltonians

$$h_1 = \int \frac{1}{8}(2u^3 - u_x^2)\mathrm{d}x$$

$$h_2 = \int \frac{1}{32}(5u^4 - 10uu_x^2 + u_{xx}^2)\mathrm{d}x$$

$$h_3 = \int \frac{1}{128}(14u^5 - 70u^2 u_x^2 + 14uu_{xx}^2 - u_{xxx}^2)\mathrm{d}x$$

and flows

$$\partial_1 u = Ru_x = \frac{1}{4}(\partial_x^3 u + 6uu_x)$$

$$\partial_2 u = R^2 u_x = \frac{1}{16}(\partial_x^5 u + 20u_x u_{xx} + 10uu_{xxx} + 30u^2 u_x)$$

$$\partial_3 u = R^3 u_x = \frac{1}{64}(\partial_x^7 u + 42u_x u_{xxxx} + \cdots)$$

We will not prove this, but instead show that ASDYM has a recursion operator and that the KdV equation is a symmetry reduction of the ASDYM equations. The ASDYM recursion operator will then reduce to the above for KdV. Other reductions of ASDYM will similarly inherit this structure. In order to do this we will need to introduce two potential formulations for the ASDYM equations.

Potential forms of the ASDYM equations

The Lax pair is a family of integrability conditions and these can be used in order to simplify the dependent variables from the four connection matrices down to one potential matrix. Such potentials are important both for making contact with hermitian geometry and with Lagrangian/Hamiltonian formulations. There are two prominent such potentials, which we will refer to as K-matrices and J-matrices respectively. They are both obtained by use of the ASDYM equations in the above coordinates:

$$[D_{\tilde{w}}, D_{\tilde{z}}] = 0, \quad [D_w, D_{\tilde{w}}] - [D_z, D_{\tilde{z}}] = 0, \quad [D_z, D_w] = 0. \quad (4.2)$$

The derivation of both forms starts with the observation that the first of these implies that there exists a frame for the bundle (a gauge) whose basis vectors are constant along $D_{\tilde{w}}$ and $D_{\tilde{z}}$ so that $D_{\tilde{w}} = \partial_{\tilde{w}}$ and $D_{\tilde{z}} = \partial_{\tilde{z}}$, i.e. $A_{\tilde{z}} = A_{\tilde{w}} = 0$. When the coordinates are complex, this shows that the connection is compatible with the complex structure and admits holomorphic sections which yield a holomorphic frame.

The J-matrix: the last of equations (4.2) is the integrability condition for the existence of a second frame that is constant along D_z and D_w, and so it will represented relative to the first by a matrix function J such that

$$(\partial_z + A_z)J = (\partial_w + A_w)J = 0, \quad \text{so that } A_z = -J_z J^{-1}, \quad A_w = -J_w J^{-1}.$$

This determines J up to $J \to \tilde{h}(z, w)Jh(\tilde{z}, \tilde{w})$, the function $\tilde{h}$ arising from the freedom in the choice of holomorphic frame. The final equation yields

$$\partial_{\tilde{z}}(J_z J^{-1}) - \partial_{\tilde{w}}(J_w J^{-1}) = 0. \quad (4.3)$$

If we choose a reality structure in which the above complex coordinates yield a Kahler metric, and choose unitary gauge group, then J can be chosen to be the hermitian metric expressed in a holomorphic frame and there is a standard expression for the curvature in this frame, $F = \bar{\partial}((\partial J)J^{-1})$ where $d = \partial + \bar{\partial}$ is the usual decomposition of d into its holomorphic and anti-holomorphic parts. Then (4.3) is just $\omega \wedge F = 0$. This is a natural equation for a hermitian metric on a holomorphic vector bundle over a Kahler manifold in 4 dimensions.

The K-matrix: to obtain the the K-matrix potential we start with the gauge in which $A_{\tilde{w}} = A_{\tilde{z}} = 0$ and use the second equation of (4.2) to deduce that

$$\partial_{\tilde{w}} A_w - \partial_{\tilde{z}} A_z = 0$$

which implies that there exists a potential K taking values in the (complexified) Lie algebra of the gauge group such that

$$A_w = \partial_{\tilde{z}} K , \quad \text{and} \quad A_z = \partial_{\tilde{w}} K .$$

K is unique up to

$$K \to hKh^{-1} + \tilde{z} h \partial_w h^{-1} + \tilde{w} h \partial_z h^{-1} + f(z,w) , \quad \text{where} \quad h = h(z,w).$$

The last ASDYM equation then implies that

$$(\partial_z \partial_{\tilde{z}} - \partial_w \partial_{\tilde{w}})K + [\partial_w K, \partial_z K] = 0 .$$

Lagrangians

Although there is a Lagrangian and corresponding phase space framework for the full Yang–Mills equations, that Lagrangian does not lead to a phase space framework (i.e. Hamiltonian and symplectic form) for the ASDYM equations. However, there are Lagrangians for each of the potential forms of the equations given above.

The K-matrix Lagrangian is

$$L[K] = \int_{\mathbb{R}^4} \text{tr}((\partial_a K)(\partial^a K) + \frac{1}{3} K[\partial_z K, \partial_w K]) \, d\tilde{w} d\tilde{z} dw dz .$$

The J-matrix Lagrangian doesnt have such a straightforward formula. The simplest, due to Donaldson, is

$$L[J] = \int x^a x_a \text{tr}(F \wedge F) .$$

It can also be written using a 'Wess–Zumino–Witten (WZW)' term, or by expressing $J = UL^{-1}$ where U is upper triangular with 1's down the diagonal and L is lower triangular. Then

$$S[J] = \frac{1}{2} \int \text{tr}(2U^{-1} \bar{\partial} U \wedge L^{-1} \partial L - L^{-1} \partial L \wedge L^{-1} \bar{\partial} L) \wedge \omega .$$

In order to develop the theory, we only need to know that

$$\delta S = -\int \mathrm{tr}(J^{-1}\delta J)\partial(J^{-1}\bar{\partial}J) \wedge \omega\,.$$

The Hamiltonian formulation

Neither $\mathbb{E}$ or $\mathbb{U}$ signature gives rise to an evolution equation equation although one cast the equations as evolution equations in the complex. Nevertheless, one can use the two Lagrangians to obtain two formal expressions for the symplectic structures and Hamiltonians. These give rise to different symplectic structures which are compatible in the sense required of a recursion operator.

Formally, these Lagrangians endow the phase space

$$M = \{\text{ Space of solutions to ASDYM eqs.}\}/\{\text{ gauge }\}$$

with a symplectic structure (here we are being sloppy about the residual gauge freedom in the potential forms of the equations which must be factored out to obtain the space of solutions to the ASDYM equations — see Mason & Woodhouse (1996) for a more careful treatment).

The tangent space to M at a given solution $A \in M$ is the space of solutions δA to $D\delta A = -{}^*D\delta A$. It can be identified with perturbations $J^{-1}\delta J$ to the J- or K-matrix equations. These satisfy

$$D^a D_a \delta K = 0, \quad D^a D_a J^{-1}\delta J = 0$$

and so both potential forms identify $T_A M$ with

$$\mathcal{W} = \{\text{ space of solutions } \phi \text{ to } D_a D^a \phi = 0 \}\,.$$

The symplectic structure in both cases is

$$\Omega(\phi_1, \phi_2) = \int \{\phi_1{}^* D\phi_2 - \phi_2{}^* D\phi_1\}\,.$$

One can check that it is closed etc.

However, J and K give two different maps $j, k : \mathcal{W} \to T_A M$ with

$$j(\phi) = D_{\tilde{w}}\phi\,\mathrm{d}\tilde{w} + D_{\tilde{z}}\phi\,\mathrm{d}\tilde{z}\,, \quad k(\phi) = D_z\phi\,\mathrm{d}\tilde{w} + D_w\phi\,\mathrm{d}\tilde{z}\,.$$

These maps can be used to push forward Ω to give two different symplectic forms on $T_A M$. Alternatively it defines the symplectic form and a recursion operator $R = k \circ j^{-1}$.

Either Lagrangian can be used to obtain the Hamiltonians that generate space-time translations, but the K-matrix version is easiest to use and gives

$$H_V = \int V^b \mathrm{tr}\{\partial^a K \partial_b K - \frac{1}{2}\delta^a_b \partial^c K \partial_c K + \frac{4}{3}\alpha^{ac}K[\partial_b K, \partial_c K]$$
$$- \frac{4}{3}\alpha^{cd}\delta^a_b K \partial_c K \partial_d K)\mathrm{d}^3 x_a\},$$

where $\alpha_{ab}\mathrm{d}x^a \mathrm{d}x^b = \mathrm{d}w \wedge \mathrm{d}z$, for the Hamiltonian generating flows along the vector $V = V^a \partial_a$,

Proposition 4 *The symplectic structure above and R together endow the space of solutions to ASDYM with a recursion operator for which the flows along the coordinate axes are biHamiltonian.*

Sketch proof: note that identifying $\mathcal{W}$ with $T_A M$ using k gives the recursion operator (also denoted R) $R = j^{-1}k$ on $\mathcal{W}$

$$\phi' = R\phi \quad \Leftrightarrow \quad D_{A0}\phi = D_{A1}\phi'.$$

To prove the result we need to show first that $\Omega(R\phi, \phi') = \Omega(\phi, R\phi')$. This follows from an integration by parts (exercise). From this it follows that the forms $\Omega_k(\phi, \phi') = \Omega(R^k\phi, \phi')$, $k \in \mathbb{Z}$ are skew. One must check a further condition to show that these forms are closed, see Mason and Woodhouse (1996) for full details.

ASDYM hierarchies and generalizations

As before for the KdV equation, $\partial_a K$ defines a linearized solution on which R can act and so we can use $R^i \partial_a(K)$ to define flows of a hierarchy. Write, for $a = 0, 1$, $i = 0, 1$,

$$x^{Ai} = \begin{pmatrix} \tilde{z} & w \\ \tilde{w} & z \end{pmatrix}.$$

We discover that the field equations are equivalent to

$$\frac{\partial K}{\partial x^{A1}} = R\left(\frac{\partial K}{\partial x^{A0}}\right)$$

(exercise). We can now define higher flows by setting

$$\frac{\partial K}{\partial x^{A,i+1}} = R\left(\frac{\partial K}{\partial x^{Ai}}\right).$$

From the definition of the recursion operator this can be recast as the equations on K

$$\frac{\partial K}{\partial x^{A,i+1}} = R\left(\frac{\partial K}{\partial x^{Ai}}\right) \quad \text{which implies} \quad D_{B0}\frac{\partial K}{\partial x^{A,i+1}} = D_{B1}\left(\frac{\partial K}{\partial x^{Ai}}\right).$$

These equations have the Lax system

$$L_{Ai} = \partial_{Ai} + A_{Ai} - \lambda \partial_{A,i-1} \quad \text{where } A_{Ai} = \partial_{A,i-1} K.$$

These equations are a hierarchy for the ASDYM equations.

The above formulation of the ASDYM hierarchy arose from a special gauge choice. To remove this, we can consider the Lax system

$$L_{Ai} = \partial_{Ai} + A_{Ai} - \lambda(\partial_{A,i-1} + \tilde{A}_{A,i-1})$$

which can be reduced to the first by using the commutativity conditions. Note that i can be continued to negative values of i also. The condition that this Lax system commutes, then, gives the field equations of the ASDYM hierarchy.

A final generalization is to let the index A extend from $1, \ldots, N$ also. This gives the *generalised ASDYM hierarchy*. This is particularly natural when i ranges over just 2 values and $N = 2k$ as the Lax system arises from a connection on a $4k$-dimensional space and with appropriate reality conditions determines a hypercomplex connection on bundles on $\mathbb{H}^k$, i.e. a connection that is holomorphic with respect to the three complex structures of a quaternionic Kahler manifold. Thus this system defines a hierarchy for the equations on a connection compatible with a hypercomplex structure (a specification of three complex structures (I, J, K) that satisfy the quaternion relations).

4.4 Reductions of the ASDYM equations

The easiest way to spot reductions is simply by comparison of the Lax Pair of the desired system with that of the ASDYM equations; one needs only to express the Lax pair in a form that is linear in λ with appropriate derivatives and matrices. (As an exercise the reader might like to reformulate the Lax pair for the Euler top so that it has this property.) This is effective but one would often like to do more. One would like to understand the precise natural conditions on the ASDYM field so that it reduces to the required sytem. One would also like to be able to express the dependent variables of the reduced system as gauge invariants that determine the full connection so that there is a 1-1 correspondence between gauge equivalence classes of solutions to the ASDYM equations and the space of solutions to the integrable system in question. Also, there are some surprises upon symmetry reduction. One discovers that the reduced equations often have additional symmetries, sometimes infinite-dimensional. Sometimes two quite different routes give the same equation.

Reductions are classified by the ingredients involved, i.e.

- A choice of gauge group.
- A choice of subgroup H of the conformal group and action on the bundle.
- A choice of reality structure.
- A choice of potential formulation or gauge.
- A choice of certain constants of integration that arise in the reduction process.

For each generator h of H one must have an action $\mathcal{L}_h$ on the bundle satisfying $\mathcal{L}_h(fs) = h(f)s + f\mathcal{L}_h s$. This gives rise to a natural endomorphism of the bundle, $\Phi_h = D_h - \mathcal{L}_h \in \Gamma(\text{End}(E))$. We refer to Φ_h as a 'Higgs field' since this is precisely how Higgs fields arise in Kaluza–Klein theory. An invariant gauge is one in which the frame is Lie derived along h, so $\mathcal{L}_h = h$ and $\Phi_h = A(h)$ (cf. the discussion of monopoles).

Many examples are described in Mason & Woodhouse (1996). Here we will examine the reduction by a null translation, and then a further translation to give the KdV and nonlinear Schrodinger (NLS) equations. Finally we consider the Ernst equations. These all illustrate interesting features. The case of the Painlevé equations will be the subject of the notes in this volume by N. M. J. Woodhouse. We will consider only gauge groups that are real forms of $SL(2, \mathbb{C})$.

Reduction by a null translation

Let us choose gauge group $SL(2, \mathbb{R})$, the totally real coordinates on $\mathbb{U}$ and symmetry in the $\partial_{\tilde{z}}$ direction with associated Higgs field Φ. Note that in general one of the ASDYM equations reduces to the condition $D_{\tilde{w}}\Phi = 0$ and this implies that the invariants $\text{tr}(\Phi^r)$ are functions of (z, w) alone. These are constants of integrations; the remaining equations are deterministic on the remaining variables.

Remark. The K-matrix equation is

$$\partial_w \partial_{\tilde{w}} K + [\partial_z K, \partial_w K] = 0$$

and this equation exhibits a new larger coordinate freedom, $(w, z, \tilde{w}) \rightarrow (w', z', \tilde{w}') = (f(w, z), z, \tilde{w})$ for arbitrary f, (use $(\partial_{w'}, \partial_{z'}, \partial_{\tilde{w}'}) =$

$(f_w^{-1}\partial_w, \partial_z - f_z f_w^{-1}\partial_w, \partial_{\tilde{w}}))$. Under this transformation $\Phi = K_w \to \Phi/f_w$.

To reduce the equations, choose an invariant gauge such that $A_{\tilde{w}} = 0$ also. The residual gauge freedom is now given by $g = g(z, w)$. We then find the reduced equations

$$\partial_{\tilde{w}}\Phi = 0, \quad D_z\Phi + [\partial_{\tilde{w}}, D_w] = 0, \quad [D_z, D_w] = 0.$$

The first equation implies that Φ is independent of $\tilde{w}$. The residual gauge freedom can be used to reduce Φ to normal form, and there are two nontrivial cases.

(i) $\operatorname{tr}\Phi^2 = 0$ The normal form here is $\Phi = \begin{pmatrix} 0 & 0 \\ 1 & 0 \end{pmatrix}$ with residual gauge freedom $g = \begin{pmatrix} 1 & 0 \\ h(w, z) & 1 \end{pmatrix}$. If we set $A_w = \begin{pmatrix} q & p \\ r & -p \end{pmatrix}$, we dicover that $\partial_{\tilde{w}}p = 0$. Thus p is another constant of integration and can be chosen to be -1 (or alternatively we can, when $p \neq 0$ reduce p to -1 by means of the coordinate freedom). We then obtain $\partial_{\tilde{w}}(r + \partial_w q + q^2) = 0$ and we can use the residual gauge freedom to set $r + \partial_w q + q^2 = 0$. With this, and further work, we discover that all entries of A_w and A_z are determined in terms of q, and

$$4q_{wz} - q_{www\tilde{w}} - 8q_w q_{w\tilde{w}} - 4q_{\tilde{w}} q_{w\tilde{w}} = 0. \tag{4.4}$$

The residual gauge freedom now yields $q \to q + a(z)$.

(ii) For $\operatorname{tr}(\Phi^2) \neq 0$, as above, we can either consider $\operatorname{tr}(\Phi^2)$ as a constant of integration and choose it to be -2 in order to obtain autonomous equations, or we can perform the coordinate transformation above to reduce it to that value. Then the gauge can be used to reduce Φ to $\Phi = \operatorname{diag}(i, -i)$ with residual gauge freedom consisting of diagonal matrices. This can be used to reduce the diagonal entries of A_w to 0, so that $A_w = \begin{pmatrix} 0 & q \\ p & 0 \end{pmatrix}$ for some p amd q. The equations imply that $A_z = \frac{1}{4}\begin{pmatrix} -V & 2q_{\tilde{w}} \\ -2p_{\tilde{w}} & V \end{pmatrix}$ for some V and

$$V_w = -2(pq)_{\tilde{w}}, \quad 2q_z = q_{w\tilde{w}} + qV, \quad 2p_z = -p_{w\tilde{w}} - pV.$$

This has real form, starting from an $SU(2)$ connection on $\mathbb{U}$, and replacing $(z, w, \tilde{w})$ by (t, x, y),

$$i\psi_t = \psi_{xy} + V\psi, \quad V_x = 2(|\psi|^2)_y. \tag{4.5}$$

These are degenerate versions of the KP and Davey–Stewartson equations respectively. (The full forms of the KP and Davey–Stewartson equations are unlikely to be reductions of the Yang–Mills equations with a finite dimensional gauge group, see Mason (1995) for a more extended discussion, although they do appear in Ablowitz & Clarkson (1991) as reductions from ASDYM with an infinite dimensional gauge group.)

With a further symmetry along $\partial_w - \partial_{\tilde{w}}$ it can be readily seen that these reductions give the non-linear Schrodinger and KdV equations respectively. Putting $x = y \ (= w = \tilde{w})$, then for KdV, we put $u = q_x$ and then (4.4) reduces to the KdV equation, $4u_t - u_{xxx} - 12uu_x = 0$. For the nonlinear Schrodinger equation, it is clear from (4.5) that with $x = y$, $V = |\psi|^2 + f(t)$ for some $f(t)$ and this f can be reduced to zero by a residual gauge transformation yielding the non-linear Schrodinger equation (NLS), $i\psi_t = \psi_{xx} + |\psi|^2\psi$. These reductions extend to reductions of the hierarchy of the ASDYM equations to that of KdV and NLS.

Alternatively, with different gauge choices, and a further symmetry along ∂_z, we obtain the Sine–Gordon equation and nonlinear σ models with values in $SU(2)$ and Hitchin's Higgs bundle equation. This latter system now has *conformal* invariance in 2-dim, $(w, \bar{w}) \rightarrow (f(w), \overline{f(w)})$ whose origins can already be seen in the infinite dimensional symmetry that arises after the null translation reduction above. Higher rank gauge groups yield systems such as the Toda field theory and harmonic maps into symmetric spaces. Reduction by a further translation yields the Euler spinning top equations and large families of generalizations.

Non-autonomous reductions

If we impose symmetries that are not translations, the reduced equations will generally no longer be translation invariant. With one symmetry around a rotation in a 2-plane in $\mathbb{E}$, one obtains the Bogomolny equations on hyperbolic 3-space, i.e. the equations $F_A = {}^* D_A \Phi$ on a connection A and Higgs field Φ where now the $*$ is that of hyperbolic 3-space. With a further symmetry and gauge group $SL(2, \mathbb{C})$ one obtains the Ernst equations, the Einstein vacuum equations from general relativity with two commuting symmetries and a 2-surface orthogonality condition, and with a 3rd symmetry (still with $SL(2, \mathbb{C})$ gauge group, we obtain the Painlevé equations.

The Ernst equations

We first discuss the Ernst equations in a form due to Ward (1983). We consider a metric on a space-time with two commuting symmetries such that the 2-planes orthogonal to the symmetries are integrable. If we introduce coordinates (x, r, y^i) such that $\partial/\partial y^i$ are the symmetries, the metric can be put in the form

$$ds^2 = J_{ij} dy^i dy^j - \Omega^2 (dx^2 + dr^2) \quad \text{where} \quad \det J = -r^2 \,.$$

The Einstein vacuum equations then reduce to

$$\partial_x (rJ^{-1}\partial_x J) + \partial_r (rJ^{-1}\partial_r J) = 0 \,,$$
$$2\partial_\xi (\log r\Omega^2) = ir\text{tr}((J^{-1}\partial_\xi J)^2) \,, \quad \xi = x + ir \,.$$

The equation for Ω is linear and integrable (it is overdetermined) as a consequence of the equation for J, and so one usually focuses on the first equation.

This can be reduced from Yang's J-matrix equation version of the ASDYM equations by setting $z = x + it$ and $w = re^{i\theta}$ and imposing symmetries along ∂_t and ∂_θ. It is easily seen that the equation then reduces to the above.

It can be shown that a version of the Backlund transformation introduced earlier gives rise again to this same reduction of Yang's equation, but now in terms of the twist and magnitude of one of the Killing vectors, and this is in fact Ernst's original form of these equations. The interplay between these two representations is nontrivial and can be used to build the Geroch group of transformations of solutions to these equations (a gauge transformation of one does not lead to a gauge transformation of the other).

The Painlevé equations

The Painlevé equations arose from Painlevé's classification of all second order ODEs that satisfy the Painlevé property that solutions admit no movable branching or essential singularities. This led to a list of six new equations that could not be reduced to known equations and whose solutions define new transcendental functions. It turns out that there are 5 distinct abelian 3-dimensional subgroups of the conformal group that satisfy a certain non-degeneracy property. Reduction of $SL(2, \mathbb{C})$ ASDYM by the first of these yields Painlevé's first and second equations, according to a choice of normal form of a null Higgs field. Each of the remaining four Painlevé equations correspond to each of the remaining 4 groups. For a full discussion see Mason & Woodhouse (1996).

4.5 Further topics

Clearly many more equations can be obtained by considering other choices of symmetries and gauge groups. The aim in Mason & Woodhouse (1996) is to show how familiar examples arise and are related to each other and to explore some related equations that are obtained when more general choices are obtained rather than develop a full classification. A full classification is likely to be rather unwieldy as, for example, there are more than 50 distinct (families of) 2-dimensional subgroups of the conformal group, and, for higher dimensional Lie groups, the classification of conjugacy classes of elements of the Lie algebra, as required for the choices of normal forms of Higgs fields, is already complicated.

Another related programme is to consider reductions of the anti-self-dual vacuum Einstein equations (ASDVE equations) on a metric. Such metrics have vanishing Ricci tensor and anti-self-dual Wey tensor. In Euclidean signature this condition is equivalent to the conditions that there are three Kahler structures, (I, J, K), satisfying the quaternion relations and so such metrics are known as hyperkahler. These equations are also integrable in the sense that there is a Lax pair and twistor construction. The equations can be expressed as a reduction of ASDYM, but only if infinite-dimensional gauge groups (diffeomorphism groups) are used, see Mason & Newman (1989), Ward (1990). The scheme of reductions of these equations with one and two symmetries is now reasonably well understood.

A spinoff from understanding the various chains of reductions is that by knowing solutions to a reduction, one can deduce solutions to the original equation, so for example, by knowing that the Painlevé equations are reductions of the Ernst equations, one can find solutions to the full Einstein vacuum equations in terms of the Painlevé transcendents, Calvert & Woodhouse (1997).

The most important corollary of the the realization of integrable systems as reductions of the ASDYM or ASDVE equations is the existence of twistor correspondences for integrable systems for all these equations. Indeed, many equations that are unlikely to be reductions of the ASDYM or ASDVE equations nevertheless admit twistor correspondences in some form.

The second part of Mason & Woodhouse (1996) is concerned with twistor theory. As mentioned above there are twistor constuctions for the ASDYM equations and the ASDVE equations. In the first case, solutions to the ASDYM equations are transformed via the Ward transform to holomorphic vector bundles over regions in $\mathbb{CP}^3$ (complex projective 3-space which in this context is twistor space), and in the second, solutions correspond to deformations of the regions in twistor space themselves. These constructions amount to a geometric general solution of the equations in the sense that the data on twistor space is freely prescribable (although the reconstruction of the solution on space-time is far from trivial). It emerges that one can see that much of the standard machinery of integrable systems theory arises from these constructions. In particular methods based on the Riemann–Hilbert problem such as dressing procedures and the inverse scattering transform are specializations of the twistor constructions.

It is difficult for twistor theory to improve on existing methods for differential equations with a well developed theory such as the KdV equations. The real benefit of the interactions has been in the other direction; well developed theory for equations such as the KdV equation can be applied in certain cases to the full ASDYM equations or other reductions using twistor theory. For example, the spectral curve construction for monopoles, Hitchin (1982), extends a technique for what are in effect integrable ordinary differential equations to equations in three dimensions subject to boundary conditions. Another example, is proposition 10.5.1 in Mason & Woodhouse (1996) which gives a construction for global solutions to the ASDYM equations in split signature based on the inverse scattering tranform for global solutions to the KdV equations on the line. One might hope that ideas from the theory of integrable systems will eventually provide some of the techniques that twistor theory needs to realize its more fundamental aspirations in mathematical physics.

4.6 Exercises

(i) Check that the KdV and KP Lax pairs do indeed give rise to the corresponding equations and identify the various entries in the matrices or operators in terms of the KdV potential q or the KP u.

(ii) Show that $F^{**} = \pm F$ depending on the signature.

(iii) Show that $F \rightarrow F^*$ interchanges the electric and magnetic fields with factors of ± 1 or $\pm i$ and determine the different cases in the corresponding signatures.

(iv) Find an expression for the t'Hooft ansatze solutions using ordinary space-time derivatives and Pauli matrices. Check directly that the ASDYM equations follow from the Laplacian on ϕ.

(v) Show in the case of the $k = 1$ instanton solutions using the t'Hooft ansatz that it is regular at a. Use the gauge transformation $x^{AA'}/|x|$ to show that there exists a gauge in which $A \to 0$ as $|x| \to \infty$. (If you're feeling brave, check by direct integration that the instanton number is -1.)

(vi) Make the ansatz $\Phi = f(r)\mathbf{x} \cdot \boldsymbol{\sigma}$ and $\mathbf{A} = g(r)\mathbf{x} \wedge \boldsymbol{\sigma}$ where $\boldsymbol{\sigma} = (\sigma_1, \sigma_2, \sigma_3)$ are the Pauli sigma matrices and solve the monopole equation. This should give the 1 monopole solution.

(vii) Work through the conditions arising from the solution generating ansatze on $\mathbb{U}$ with real coordinates in the case $n = 2$, $k = 1$, and with v depending only on $\omega = \mu(\mu z + \tilde{w}) + \mu w + \tilde{z} = \mu^2 z + \mu(w + \tilde{w}) + \tilde{z}$ to determine a ψ and a corresponding solution to the ASDYM equations. Put $2x = w + \tilde{w}$ and $t + y = z$, $t - y = \tilde{z}$ and show that following the ansatze through will lead to a solution to the $\mathbb{R}^{2+1}$ solution to the Bogomolny equation. Note that this solution only depends on (x, y), i.e. is static when $\mu = i$. [Hint: first try to show that

$$\psi = 1 + \frac{\mu - \bar{\mu}}{\lambda - \mu}\left(\frac{v^* \otimes v}{v \cdot v^*}\right)$$

satisfies the required conditions, where $v = v(\mu z + \tilde{w}, \mu w + \tilde{z})$. For a simple special example, put $v = (1, \omega)$.]

References

Ablowitz, M. & Clarkson, P. A. (1991). *Solitons, nonlinear evolution equations and inverse scattering*, LMS lecture notes, **149**, CUP.

Atiyah, M. F. (1979). *Geometry of Yang–Mills fields*, Academia Nazionale Dei Lincei, Scuola Normale Superiore, Pisa.

Calvert, G. & Woodhouse, N. M. J. (1997). Painlevé transcendents and Einstein's equations, *Class. & Quant. Grav.*, **13**, No 4, L33-9.

Hitchin, N. J. (1982). Monopoles and geodesics, *Comm. Math. Phys.*, **83**, 589-602.

Huggett, S. A. & Tod, K. P. (1994). *An introduction to twistor theory*, LMS student texts, **4**, CUP.

Magri, F. (1978). A simple model of the integrable Hamiltonian equation, *J. Math. Phys.*, **19**, 1156-62.

Mason, L. J. (1995). Generalized twistor correspondences, d-bar problems and the KP equations, in *Twistor Theory*, ed. S. Huggett, Lecture Notes in Pure and Applied Mathematics, vol. 169, Marcel Dekker.

Mason, L. J. & Newmam, E. T. (1989). A connection between the Einstein and Yang–Mills equations, *Comm. Math. Phys.*, **121**, 659-68.

Mason, L. J. & Woodhouse,, N. M. J. (1996). *Integrability, self-duality and twistor theory*, LMS Monographs, OUP.

Uhlenbeck, K. (1982). Removable singularities in Yang–Mills fields, *Comm. Math. Phys.*, **83**, 11-29.

Ward, R. S. (1983). Stationary axisymmetric space-times: a new approach, *Gen. Rel. Grav.*, **15**, 105-9.

Ward, R. S. (1985). Integrable and solvable systems and relations amongst them, *Phil. Trans. Royal Soc.*, **315**, 451-7.

Ward, R. S. (1990). The $SU(\infty)$ chiral model and self-dual vacuum spaces, *Class. & Quant. Grav.*, **7**, L217-22.

Ward, R. S. & Wells, R. (1990). *Twistor geometry and Field theory*, CUP.

5

Curvature and integrability for Bianchi-type IX metrics

K. P. Tod

Mathematical Institute, St Giles, Oxford OX1 3LB

Abstract

In this seminar, I review the various curvature conditions that one might wish to impose on a Bianchi-type IX metric, and the direct route to the self-dual Einstein metrics obtained from solutions of the Painlevé VI equation.

In four-dimensions there are several 'nice' conditions which one may impose on a Riemannian metric. For example, one may require it to be Kähler, or Einstein, or to have anti-self-dual (ASD) Weyl tensor. These possibilities are set out in figure 5.1, which I have used before (Tod 1995, 1997). The overlapping regions in the figure also correspond to interesting conditions: a Kähler metric with zero-scalar curvature ('scalar-flat' in the terminology of (LeBrun 1991)) has ASD Weyl tensor; ASD Einstein is known as quaternionic Kähler; ASD Ricci-flat is hyper-Kähler. Elsewhere in the top circle of the diagram are hyper-complex metrics, which have three integrable complex structures with the algebra of the (unit, complex) quaternions but are not Kähler.

Conditions on the curvature are most readily imposed using Cartan calculus, so suppose that (e_0, e_1, e_2, e_3) is a normalised basis of 1-forms in some Riemannian 4-manifold. Define a basis of SD 2-forms by

$$\phi_1 = e_0 \wedge e_1 + e_2 \wedge e_3$$
$$\phi_2 = e_0 \wedge e_2 + e_3 \wedge e_1$$
$$\phi_3 = e_0 \wedge e_3 + e_1 \wedge e_2$$

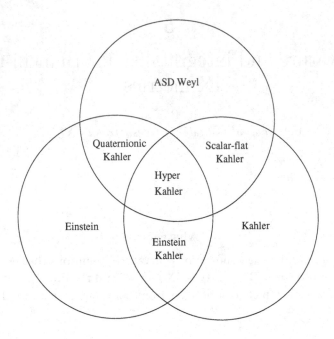

Fig. 5.1. Different field equations in 4-dimensions

so that $\star\phi_i = \phi_i$, in terms of the Hodge star or dual, and a basis of ASD 2-forms by

$$
\begin{aligned}
\psi_1 &= e_0 \wedge e_1 - e_2 \wedge e_3 \\
\psi_2 &= e_0 \wedge e_2 - e_3 \wedge e_1 \\
\psi_3 &= e_0 \wedge e_3 - e_1 \wedge e_2
\end{aligned}
$$

so that $\star\psi_i = -\psi_i$. Now define the connection 1-forms for the SD part of the Levi–Civita connection, $\alpha_{ij} = -\alpha_{ji}$, by

$$
d\phi_i = \alpha_{ij} \wedge \phi_j
$$

and the corresponding curvature 2-forms Ω_{ij} by

$$
\Omega_{ij} = d\alpha_{ij} - \alpha_{ik} \wedge \alpha_{kj}.
$$

Since the Ω_{ij} are 2-forms, we may expand them in terms of ϕ_i and ψ_i as

$$
\Omega_{ij} = W_{ijk}\phi_k + \Phi_{ijk}\psi_k
$$

in terms of two sets of coefficients, W_{ijk} and Φ_{ijk}. The first Bianchi identity is

$$\Omega_{ij} \wedge \phi_j = 0$$

which implies that $W_{ijj} = 0$. It follows that W_{ijk} has 6 independent components, 5 corresponding to the SD Weyl tensor and one, the totally-anti-symmetric part, corresponding to the Ricci scalar, while Φ_{ijk} has 9 components corresponding to the trace-free Ricci tensor.

Various field equations can now be imposed as, for example:

(i) ASD Weyl tensor iff $W_{ijk} = \Lambda \epsilon_{ijk}$ where Λ is a multiple of the Ricci scalar (since this condition on the Weyl tensor is conformally-invariant, it is often convenient to choose the conformal scale so that Λ vanishes).

(ii) ASD Einstein iff $W_{ijk} = \Lambda \epsilon_{ijk}$ and $\Phi_{ijk} = 0$.

(iii) Hyper-Kähler iff $W_{ijk} = 0 = \Phi_{ijk}$.

From now on, we shall be interested just in Bianchi-type IX metrics, that is metrics with an action of $SU(2)$ transitive on hypersurfaces. Such a metric can be written in terms of the basis $(\sigma_1, \sigma_2, \sigma_3)$ of left-invariant 1-forms on $SU(2)$ in the form

$$ds^2 = w_1 w_2 w_3 dt^2 + \frac{w_2 w_3}{w_1} d\sigma_1^2 + \frac{w_3 w_1}{w_2} d\sigma_2^2 + \frac{w_1 w_2}{w_3} d\sigma_3^2 \qquad (5.1)$$

where the w_i are three functions of t, and the σ_i satisfy

$$d\sigma_1 = \sigma_2 \wedge \sigma_3$$

and cyclic permutations of this.

The virtue of writing the metric in the form of (1) is that the basis of SD 2-forms can be taken to be

$$\phi_1 = w_2 w_3 dt \wedge \sigma_1 + w_1 \sigma_2 \wedge \sigma_3$$

and cyclic permutations of this. Then it turns out that the connection 1-forms α_{ij} can be written in terms of three more functions a_1, a_2, a_3 of t as

$$\alpha_{12} = \frac{a_3}{w_3} \sigma_3$$

and cyclic permutations of this, where the a_i are determined by the equations

$$\frac{dw_1}{dt} = w_2 w_3 - w_1(a_2 + a_3)$$
$$\frac{dw_2}{dt} = w_3 w_1 - w_2(a_3 + a_1)$$
$$\frac{dw_3}{dt} = w_1 w_2 - w_3(a_1 + a_2)$$

which we shall call the first system.

We may continue with the Cartan calculus and find the curvature 2-forms in terms of derivatives of the a_i. If we now make a choice of field equations, then we obtain a second system of first-order differential equations on the a_i. If we choose something outside the top circle in the figure then typically we arrive at equations that are not integrable (for Einstein–Kähler, (Dancer and Strachan 1994)) or even chaotic (for Einstein, see e.g.(Barrow 1982)). However field equations from inside the top circle are integrable (as is to be expected from the 'self-duality implies integrability' heuristic; see e.g. Mason 1990).

We first consider the case (i) above, the case of vanishing W_{ijk}. Then the second system turns out to be

$$\frac{da_1}{dt} = a_2 a_3 - a_1(a_2 + a_3)$$
$$\frac{da_2}{dt} = a_3 a_1 - a_2(a_3 + a_1) \tag{5.2}$$
$$\frac{da_3}{dt} = a_1 a_2 - a_3(a_1 + a_2)$$

This attractive system is widely known as the Chazy system (Chazy 1910, Ablowitz and Clarkson 1991), although it was studied earlier by Brioschi (1881). We shall solve it below. Before that, we note a special solution: if we insist that all the a_i are constant in time then without loss of generality two, say a_2 and a_3, must be zero. With a_1 non-zero, one can reduce the first system to a special case of the Painlevé-III equation (Tod 1991). Also the form ϕ_1 is covariant constant in this case so that this is a scalar-flat Kähler solution, the one first found by Pedersen and Poon (1990).

We have found the solution which in figure 5.1 lies in the intersection on the right. The solutions in the triple intersection, the hyper-Kähler metrics, are characterised now by vanishing Φ_{ijk}. There are two classes. In the first, all a_i are zero, which certainly satisfies the second system, and then the first system is solved by elliptic functions. These solutions were found by Belinski et al (1979). In the second, for each i, $a_i = w_i$, so that the first system reduces to the second which we shall solve below. These solutions were found by Atiyah and Hitchin (1985, 1988).

For the general solution of (2) we follow Brioschi and introduce a new dependent variable x by

$$x = \frac{a_1 - a_2}{a_3 - a_2}$$

It is now straightforward to obtain a single third-order equation for x

$$\frac{d^3x}{dt^3} = \frac{3}{2}\frac{(\frac{d^2x}{dt^2})^2}{\frac{dx}{dt}} - \frac{1}{2}\left(\frac{dx}{dt}\right)^3\left(\frac{1}{x^2} + \frac{1}{x(x-1)} + \frac{1}{(x-1)^2}\right)$$

and this equation is solved by x equal to the reciprocal of the elliptic modular function. Now the elliptic modular function has a natural boundary in the t-plane, so that the a_i and hence also the w_i have a natural boundary in the t-plane, whose location depends on the constants of integration i.e. a movable natural boundary. This suggests that these self-duality equations are not integrable despite what was said above (integrability in the context of ODEs is usually regarded as being equivalent to the Painlevé property, that all movable singularities are poles). We shall see the resolution of this puzzle below. Meanwhile, we return to the question of solving the first system.

We introduce new dependent variables $\Omega_1, \Omega_2, \Omega_3$ according to

$$w_1 = \frac{1}{\sqrt{x(1-x)}}\frac{dx}{dt}\Omega_1$$

$$w_2 = \frac{1}{x\sqrt{(1-x)}}\frac{dx}{dt}\Omega_2$$

$$w_3 = \frac{1}{(1-x)\sqrt{x}}\frac{dx}{dt}\Omega_3$$

and switch independent variables from t to x. The first system becomes

$$\frac{d\Omega_1}{dx} = \frac{\Omega_2\Omega_3}{x(1-x)}$$

$$\frac{d\Omega_2}{dx} = \frac{\Omega_3\Omega_1}{x} \tag{5.3}$$

$$\frac{d\Omega_3}{dx} = \frac{\Omega_1\Omega_2}{(1-x)}$$

which has the first integral

$$\gamma = \frac{1}{2}(-\Omega_1^2 + \Omega_2^2 + \Omega_3^2)$$

while the metric becomes

$$ds^2 = \Theta\left(\frac{dx^2}{x(1-x)} + \frac{\sigma_1^2}{\Omega_1^2} + \frac{(1-x)\sigma_2^2}{\Omega_2^2} + \frac{x\sigma_1^2}{\Omega_3^2}\right) \tag{5.4}$$

where

$$\Theta = \frac{\Omega_1\Omega_2\Omega_3}{x(1-x)}\frac{dx}{dt}.$$

(The system (3) with different derivations and motivations, has been studied by Fokas et al (1986) and by Dubrovin (1990).)

To solve the new version, (3), of the first system, we seek an equation for Ω_3 alone. Because of the existence of the first-integral γ this will be second-order. To make it recognisable, we introduce a new independent variable z by

$$x = \frac{4\sqrt{z}}{(1+\sqrt{z})^2}$$

and a new dependent variable V by

$$\Omega_3 = \frac{z}{V}\frac{dV}{dz} - \frac{V}{2(z-1)} - \frac{1}{2} + \frac{z}{2V(z-1)}.$$

Now we find that V satisfies the Painlevé-VI equation with the parameters $(\alpha, \beta, \gamma, \delta)$ in the notation of Ince (1956) or Ablowitz and Clarkson (1991), equal to $(\frac{1}{8}, -\frac{1}{8}, \gamma, \frac{1}{2}(1-2\gamma))$. (The reduction of the system (3) to Painlevé-VI was also found by Chakravarty (1993)). Thus as anticipated the equations for the conformal structure have the Painlevé property (being in fact reducible precisely to one of the Painlevé equations!) but it is the conformal factor in (4) which contains the function $x(t)$, which has the natural boundary. This choice of conformal factor can be thought of as just a gauge choice, to make the Ricci scalar vanish.

At this point, we have found the general solution of the form (1) inside the top circle in the figure. (Note that there are ASD Bianchi-type IX metrics which are not diagonal in the chosen invariant basis of 1-forms; see (Maszczyk et al 1993)). Now we ask: can we change the conformal factor Θ in order to make these metrics Einstein? This would give the general metric in the left-hand intersection in the figure (which must be diagonal in this basis by a general argument, see e.g. (Tod 1995)).

This question can be answered by a brute-force calculation: simply write down the desired condition as a set of equations on Θ and try to solve. We find (Tod 1995) that this is only possible if $\gamma = 1/8$ and that then the solution may be written as

$$\Theta = \frac{N}{D^2}$$

where

$$
\begin{aligned}
N &= 2\Omega_1\Omega_2\Omega_3(4x\Omega_1\Omega_2\Omega_3 + P) \\
P &= x(\Omega_1^2 + \Omega_2^2) - (1 - 4\Omega_3^2)(\Omega_2^2 - (1 - x)\Omega_1^2) \\
D &= x\Omega_1\Omega_2 + 2\Omega_3(\Omega_2^2 - (1 - x)\Omega_1^2).
\end{aligned}
$$

Since the equation for Ω_3 was second-order, the metric depends on two arbitrary constants. Despite the complexity of these expressions, a good deal can be said about the metrics (Hitchin 1995). In particular, with appropriate choices, these are ASD Einstein metrics on the 4-ball which fill-in the general left-invariant metric on the 3-sphere in just the way that the 4-dimensional hyperbolic metric fills in the round metric on S^3.

Acknowledgements. I am grateful to Professor Nutku and TÜBITAK for the invitation to visit the Fesa Gürsey Institute and their generous support and hospitality during the Research Semester on Geometry and Integrability.

References

M. J. Ablowitz and P. A. Clarkson 1991 *Solitons, nonlinear evolution equations and inverse scattering* LMS Lecture Note Series **149**

M. F. Atiyah and N. J. Hitchin 1985 *Phys. Lett.* **A107** 21

M. F. Atiyah and N. J. Hitchin 1988 *The geometry and dynamics of magnetic monopoles* Princeton University Press

J. D. Barrow 1982 *Gen. Rel. Grav.* **14** 523-530

V. A. Belinski, G. W. Gibbons, D. W. Page and C. N. Pope 1979 *Phys. Lett.* **B76** 433

F. Brioschi 1881 *C. R. Acad. Sci.* tXCII 1389

S. Chakravarty 1993 private communication

J. Chazy 1910 *C. R. Acad. Sci.* **150** 456

A. S. Dancer and I. A. B. Strachan 1994 *Math. Proc. Camb. Phil. Soc.* **115** 513-525

B. A. Dubrovin 1990 *Funct. Anal. Applns.* **24** 280

A. S. Fokas, R. A. Leo, L. Martina and G. Soliani 1986 *Phys. Lett.* **A115** 329

N. J. Hitchin 1995 *J. Diff. Geom.* **42** 30-112

E. L. Ince 1956 *Ordinary differential equations* Dover reprint

C. R. LeBrun 1991 *J. Diff. Geom.* **34** 223-253

L. J. Mason 1990 in *Further Advances in Twistor Theory* eds. L. J. Mason, L. P. Hughston and P. Z. Kobak, Pitman

R. Maszczyk, L. J. Mason and N. M. J. Woodhouse 1993 *Class. Quant. Grav.* **11** 65-71

H. Pedersen and Y.-S. Poon 1990 *Class. Quant. Grav.* **7** 1707

K. P. Tod 1991 *Class. Quant. Grav.* **8** 1049-1051

K. P. Tod 1994 *Phys. Lett.* **A190** 221-224

K. P. Tod 1995 in *Twistor Theory* ed S. Huggett, Dekker Lecture notes in pure and applied mathematics **169**

K. P. Tod 1997 in *Geometry and Physics* eds. J. E. Andersen, J. Dupont, H. Pedersen and A. Swann, Dekker Lecture notes in pure and applied mathematics **184**

6

Twistor theory for integrable systems

N. M. J. Woodhouse,

The Mathematical Institute, 24-29 St Giles, Oxford OX1 3LB, UK
nwoodh@maths.ox.ac.uk

Abstract

Integrable systems that arise as reductions of the anti-self-dual Yang-Mills equations have a twistor correspondence by reduction of the Ward correspondence for the anti-self-dual Yang-Mills equations themselves. These lecture notes begin by reviewing the twistor correspondence, first of all for linear fields and anti-self-dual Yang-Mills fields. They then move onto the reduced twistor correspondences for reductions of the anti-self-dual Yang-Mills equations with particular attention paid to the special examples of the Korteweg–de Vries equations and the Painlevé equations.

6.1 Lecture 1

6.1.1 Introduction and background

In two dimensions, one can solve Laplace's equation

$$u_{xx} + u_{yy} = 0$$

by introducing a complex coordinate $w = x + iy$. The equation then becomes

$$\frac{\partial^2 u}{\partial w \partial \overline{w}} = 0, \qquad (6.1)$$

which is satisfied by $u = f(w)$ for any holomorphic function f.

How many different ways can one do this? That is, in how many different ways can one introduce a complex linear coordinates on $\mathbb{R}^2$ so that Laplace's equation reduces to (6.1)? It is not hard to see that the only freedom is to replace w by aw for some constant $a \neq 0$, or to replace w by $x - iy$. The first gives nothing new since holomorphic functions of

97

w are also holomorphic functions of aw; the second tells us that $f(\overline{w})$ is also a solution.

In four dimensions the situation is more interesting. If we start with

$$u_{xx} + u_{yy} + u_{yy} + u_{tt} = 0,$$

then we can reduce it to the form

$$\frac{\partial^2 u}{\partial w^1 \partial \overline{w}^1} + \frac{\partial^2 u}{\partial w^2 \partial \overline{w}^2} = 0 \tag{6.2}$$

by putting $w^1 = x + iy$, $w^2 = z + it$. We have solutions of the form $u = f(w^1, w^2)$ for any holomorphic function of w^1 and w^2. Again we can replace the complex coordinates by any complex linear functions of w^1 and w^2 (which gives nothing new). But in this case we have other possibilities that mix up the complex coordinates with their complex conjugates. For any constant $\alpha, \beta \in \mathbb{C}$, not both zero, the substitution

$$w^1 = \alpha(x + iy) - \beta(z - it), \qquad w^2 = \alpha(z + it) + \beta(x - iy) \tag{6.3}$$

also reduces Laplace's equation to (6.2).

Suppose that we choose instead different constants α', β'. If $\alpha/\beta = \alpha'/\beta'$, then corresponding ws are related by a complex linear transformation; but if $\alpha/\beta \neq \alpha'/\beta'$, then this is not true, and we get a new class of solutions from holomorphic functions of the new complex coordinates. To put this more formally, we have a family of complex structures on $\mathbb{R}^4$, labelled by the points of the projective line (the Riemann sphere with complex coordinate $\zeta = \alpha/\beta$).† There is also a second family that one constructs in the same way, but starting instead with $x + iy$ and $z - it$.

Now let us consider a closed 2-form $F = d\Phi$ (signature apart, we think of F as an electromagnetic field). We say that F is *anti-self-dual* if it has no $dw^1 \wedge dw^2$ or $d\overline{w}^1 \wedge d\overline{w}^2$ components for any choice of ζ (there is a more geometric definition below).

Suppose that this condition holds and fix for the moment a choice of ζ. If we write

$$\Phi = \Phi_1 \, dw^1 + \Phi_2 \, dw^2 + \Phi_{\overline{1}} \, d\overline{w}^1 + \Phi_{\overline{2}} \, d\overline{w}^2 \,,$$

then the ASD condition implies that

$$\frac{\partial \Phi_{\overline{1}}}{\partial \overline{w}^2} = \frac{\partial \Phi_{\overline{2}}}{\partial \overline{w}^1}$$

---

† In the notation of Paul Tod's lectures, w^1 and w^2 are the components of the spinor ω^A and α, β are the components of the spinor $\pi_{A'}$. The point of view here is based on that of Atiyah *et al* (1978a).

and hence that

$$\Phi_{\bar{i}} + \frac{\partial f}{\partial \overline{w}^i} = 0 \qquad i = 1, 2$$

for some function f. Written in terms of the original Cartesian coordinates, this is

$$\zeta(\partial_x + i\partial_y)f - (\partial_z - i\partial_t)f + \zeta(\Phi_x + i\Phi_y) - (\Phi_z - i\Phi_t) = 0$$
$$\zeta(\partial_z + i\partial_t)f + (\partial_x - i\partial_y) + \zeta(\Phi_z + i\Phi_t) + \beta(\Phi_x - i\Phi_y) = 0 . \quad (6.4)$$

Clearly, as α and β vary, we can find a solution f that depends holomorphically on $\zeta = \alpha/\beta$. However, we cannot find a solution that is holomorphic in ζ on the whole Riemann sphere.

What is always possible, however, is to find one solution f which is holomorphic in ζ for $\zeta \neq \infty$ and another $\tilde{f}$ which is holomorphic in ζ for $\zeta \neq 0$ (including the point at infinity). It then follows that $h = f - \tilde{f}$ is holomorphic with respect to the three variables

$$\zeta(x + iy) - (z - it), \qquad \zeta(z + it) - (x - iy), \qquad \zeta$$

(away from $\zeta = 0, \infty$) since (6.4) implies $\partial_{\overline{w}^i} h = 0$. We call h the *twistor function* of F. It encodes F since we can recover f and $\tilde{f}$, and hence Φ, by splitting the Laurent series of

$$h\big(\zeta(x + iy) - (z - it), \zeta(z + it) - (x - iy), \zeta\big)$$

into its positive frequency part (f) and negative frequency part ($\tilde{f}$), as a function of ζ for fixed values of the Cartesian coordinates x, y, z, t.

The picture is typical of twistor theory: a system of partial differential equation (the ASD condition on F) in four independent variables (the space-time coordinates) has been replaced by a condition that a single function (h) should be holomorphic in the three variables w^1, w^2, ζ.

A number of interesting linear equations, such as the hypergeometric equation and Bessel's equation, are symmetry reductions of the ASD condition; and by imposing corresponding symmetry conditions on h, one can recover classical contour integral formulas for the solutions.†

In these lectures, I shall look at the corresponding *nonlinear* theory, where the electromagnteic field is replaced by a Yang–Mills field, and

---

† The idea of looking systematically at integrable systems as reductions of the ASD Yang–Mills equations is due to Ward (1985), who also first showed that ASD Yang–Mills equation could be solved by twistor methods (Ward 1977). Ward's approach to integrability was developed, in particular, by Mason, and is surveyed in Mason and Woodhouse (1996). Twistor methods themselves are due to Penrose, who first applied them in a nonlinear context in Penrose (1976). The twistor approach to the ASD condition was applied and extended in Atiyah and Ward (1977), Atiyah (1979), and Atiyah *et al* (1978b).

the reductions are integrable systems. We shall look in particular at the KdV equation and the sixth Painlevé equation. In the nonlinear theory, h is replaced by the transition matrix of a holomorphic vector bundle.

6.1.2 Complex manifolds

We shall make some use of the theory of complex manifolds and holomorphic vector bundles. Although we shall not go very far into the theory, it is important to get some intuitive feeling for the basic structures, and in particular for the rather subtle way in which ideas from complex analysis and differential geometry come together. In the remainder of this lecture, I shall try to convey this, without giving full definitions and rigorous proofs.

First, an n-dimensional complex manifold M is defined in the same way as a real manifold, except that the transition maps between local coordinate patches are required to be holomorphic. Thus in a neighbourhood of each point, we can introduce local holomorphic coordinates z^a ($a = 1, 2, \ldots, n$). The transition between two coordinate systems z^a and w^a is must satisfy:

- the functions $z^a(w^1, \ldots w^n)$ are holomorphic (that is, $\partial z^a / \partial \overline{w}^b = 0$);
- the Jacobian matrix $\partial z^a / \partial w^b$ is non singular on the overlap of the two coordinate patches.

One can think of this as a generalization of the notion of a real manifold, and so take over the standard apparatus of differential geometry to define holomorphic functions, vector fields, and differential forms, by simply adding the requirement that the components should depend holomorphically on the coordinates.

Alternatively, we can think of an n-dimensional complex manifold as a $2n$-dimensional real manifold with some additional structure, by writing

$$z^a = x^a + \mathrm{i} y^a \,,$$

with x^a and y^a real, and taking x^a, y^a as real coordinates. The additional structure is a real tensor field J, with one upper index and one lower index. As a linear operator on tangent vectors, it is given by

$$J\left(\frac{\partial}{\partial x^a}\right) = \frac{\partial}{\partial y^a}, \qquad J\left(\frac{\partial}{\partial y^a}\right) = -\frac{\partial}{\partial x^a} \,.$$

It has the property that $J^2 = -1$.

Conversely a given tensor field J on an even-dimensional smooth real manifold M such that $J^2 = -1$ is called an *almost complex structure*. Such a tensor has two n-dimensional eigenspaces, corresponding to the eigenvalues $\pm i$. The additional property of J that characterizes M as a complex manifold is that J should be integrable. That is, if two complex vector fields Z and Z' are eigenvector fields with

$$JZ = iZ, \qquad JZ' = iZ',$$

then $[Z, Z']$ is also an eigenvector with $J[Z, Z'] = i[Z, Z']$. When this holds, it is possible to give M the structure of a complex manifold by taking the local holomorphic coordinates to be solutions z^a to the *Cauchy–Riemann equations*

$$(X + iJX)z^a = 0 \quad \text{for every vector field } X.$$

It is not hard to prove this from the Frobenius theorem when M is analytic (see Appendix 8 of Kobayashi and Nomizu 1969); but it is a hard theorem (the Newlander–Nirenberg theorem) if one starts with smooth objects. When $n = 1$ (so that M has two dimensions as a real manifold), an integrable almost complex structure is the same thing as a conformal metric together with an orientation – since $X \mapsto JX$ determines a rotation through $\pi/2$ in a positive sense. In this case, the Newlander–Nirenberg theorem is the assertation that it is possible to find *isothermal coordinates*, in which the conformal metric is a multiple of

$$dx^2 + dy^2.$$

6.1.3 Examples

Two complex manifolds M, M' are isomorphic if they are diffeomorphic by a diffeomorphism $f : M \to M'$ that is biholomorphic – that is, f and f^{-1} are holomorphic in local holomorphic coordinates. Biholomorphic mappings are much harder to find than smooth ones because their analyticity gives them an extra rigidity: a diffeomorphism can be deformed in all sorts of ways in a neighbourhood of a point, but a biholomorphic map is determined by analytic continuation once it is known in a small neighbourhood of a point of M. Consequently it is possible for two complex manifolds to be diffeomorphic but not biholomorphically equivalent.

Example. *The torus.* Let α, β be two nonzero complex numbers such that their ratio is not real. We can then construct a one-dimennsional complex manifold M by taking the quotient of the Argand plane $\mathbb{C}$ by the equivalence relation $z \sim w$ whenever

$$z = w + n\alpha + m\beta \qquad \text{for some } n, m \in \mathbb{Z}.$$

As a smooth real manifold, M is diffeomorphic to the 2-dimensional torus, irrespective of the choice of α and β. So every choice gives the same real manifold. If however we construct M' by making a different choice α', β', then M and M' are isomorphic as one-dimensional complex manifolds if and only if we can find a holomorphic function F with nowhere vanishing derivative such that

$$F(z + \alpha) = F(z) + a\alpha' + b\beta', \qquad F(z + \beta) = F(z) + c\alpha' + d\beta'$$

for some integers a, b, c, d such that $ad - bc = \pm 1$. But then $\mathrm{d}F/\mathrm{d}z$ is a holomorphic on M. It must therefore be constant by Liouville's theorem (see Exercise 1.1). Since we are free to add a constant to F, we have without loss of generality $F(z) = kz$ for some constant k. But then

$$k\alpha = a\alpha' + b\beta', \qquad k\beta = c\alpha' + d\beta'.$$

We can conclude from this that M and M' are isomorphic if and only if $\omega = \alpha/\beta$ and $\omega' = \alpha'/\beta'$ are related by

$$\omega = \frac{a\omega' + b}{c\omega' + d}$$

for some integers a, b, c, d with $ad - bc = \pm 1$. More details of this example are given in Gunning's book *Lectures on Riemann surfaces* (Gunning 1966).

We shall be more exclusively interested in projective spaces, and in manifolds constucted from projective spaces.

Example. *The projective line.* This is the familiar Riemann sphere, which is a one-dimensional complex manifold with local coordinates defined by stereographic projection. If a point on the sphere has image $z \in \mathbb{C}$ under projection from the North pole and image $\overline{w}$ under projection from the South pole, then $w = 1/z$, and we have a holomorphic coordinate transformation. The domain of z is the whole sphere less the North pole; the domain of w is the whole sphere less the South pole. Unlike the torus, the sphere has a unique complex structure.

Example. *Projective spaces.* More generally, the n-dimensional projective space $\mathbb{C}P_n$ is the quotient of $\mathbb{C}^{n+1}$ by the equivalence relation

$$(Z^0, Z^1, \ldots, Z^n) \sim (\lambda Z^0, \lambda Z^1, \ldots, \lambda Z^n) \qquad \text{for some } \lambda \neq 0 \in \mathbb{C}.$$

The *homogeneous coordinates* Z^α label the points uniquely, up to an overall complex scaling factor. We make $\mathbb{C}P_n$ into a complex manifold by using the corresponding *inhomogeneous coordinates*. For example on the open set in which $Z^0 \neq 0$, we define z^a $(a = 1, \ldots, n)$ by

$$z^1 = Z^1/Z^0, \quad z^2 = Z^2/Z^0, \quad \ldots, \quad z^n = Z^n/Z^0.$$

When $n = 1$, the projective space is the same as the Riemann sphere since we can take $z = Z^1/Z^0$ and $w = Z^0/Z^1$ as the two local coordinates (in the southern and northern hemispheres).

6.1.4 Holomorphic vector bundles

In the same way, we can extend the notion of a smooth bundle $E \to M$ to that of a holomorphic bundle by requiring that E (the 'total space') and M (the 'base space') should be complex manifolds and that the maps in the usual definition should be holomorphic. Thus a *holomorphic vector bundle* E of rank k has local trivializations in which we identify the restriction of E to a suitable open subset $U \subset M$ with $U \times \mathbb{C}^k$. On the overlap of two local trivializations we have

$$(m, v') \sim (m, v) = \big(m, F(m)v'\big),$$

where $F : U \cap U' \to \mathrm{GL}(k, \mathbb{C})$ is holomorphic (that is, F is a non-singular holomorphic matrix-valued function of the local coordinates). We call F the *patching matrix* or *transition function*.

Of course, all the usual operations ($E \oplus E'$, $E \otimes E'$, $\ldots$) from differential geometry make sense in for holomorphic bundles. But there are two key ways in which holomorphic bundles behave differently: these we shall illustrate below.

Two holomorphic vector bundles E, E' are *isomorphic* if there is a biholormphic map $\rho : E \to E'$ which maps E_m linearly to E'_m for each $m \in M$.

One way that we can specify a holomorphic vector bundle is by giving its transition maps F_{ij} between the open sets of some open cover U_i (i in some indexing set): these are holomorphic maps matrix-valued functions on the intersections with the *cocyle property*

$$F_{ij}F_{jk}F_{ki} = 1$$

on each triple intersection $U_i \cap U_j \cap U_k$. It is not hard to see that every holomorphic vector bundle can be represented in this way. What is rather less obvious, but nonetheless true, is that every holomorphic bundle can be represented in this way for some fixed special choice of open cover.

A holomorphic section s is given by holomorphic maps $s_i : U_i \to \mathbb{C}^k$, with the transition rule $s_i = F_{ij}s_j$. We denote the space of holomorphic sections over $V \subset M$ by $\Gamma(V, E)$. A fundamental fact is that:

- if M is compact, then $\Gamma(M, E)$ is finite dimensional.

In the case of the trivial bundle over the Riemann sphere, this is a direct consequence of Liouville's theorem (since all sections in that case are constant).

The bundle is trivial, that is isomorphic to $M \times \mathbb{C}^k$, if we can find holomorphic *splitting matrices* $f_i : U_i \to \mathrm{GL}(k, \mathbb{C})$ such that

$$F_{ij} = f_i^{-1}f_j\,.$$

In this case, we can specify global sections by putting $s_i = f_i^{-1}c$ for some constant vector c.

We shall give an example below to illustrate that

- E can be trivial as a smooth bundle, but non-trivial as a holomorphic bundle.

Thus a holomorphic bundle can carry information beyond that contained in its toplogical structure. Indeed it is possible to deform a holomorphic vector bundle holomorphically into an inequivalent bundle. This fact is central to the twistor constructions.

Example. *Bundles over projective space.* If we represent the Riemann sphere as the projective line, as above, then we have the projection $\mathbb{C}^2 \to \mathbb{C}P_1$. This gives us a line bundle (i.e. a rank-one vector bundle) with total space $L = \mathbb{C}^2$. The fibre above the point with coordinate $z = Z^1/Z^0$ is the one-dimensional subspace of $\mathbb{C}^2$ spanned by (Z^0, Z^1). We define holomorphic sections e and $\tilde{e}$ over the z and w coordinate patches by taking

$$e = (1, Z^1), \qquad \tilde{e} = (Z^0, 1)\,.$$

These are related by $e = z\tilde{e}$; thus if s and $\tilde{s}$ are two functions respesenting the same section in the z and w coordinate patches, respectively, then $s = \tilde{s}/z$ and so the transition function is $F = 1/z = w$ on the annular intersection of the two patches.

A local section of L is the same thing as a function f on $\mathbb{C}^2$ homogeneous of degree -1, since we can put $s(z) = f \times (Z^0, Z^1)$ to get a well-defined point of L which depends only on the ratio $z = Z^0/Z^1$.

Since there are no global holomorphic homogeneous functions of degree -1 on $\mathbb{C}^2 - \{0\}$, there are no global sections of L, a fact that we can also see in another way: a global section would be represented by an entire function $s(z)$ with the property that $zs(z) = s(w^{-1})/w$ is holomorphic in $w = 1/z$ at $w = 0$. By expanding zs in a Laurent series, it is easy to see that this is impossible.

Because its sections are functions of degree -1, it is customary to denote L by $\mathcal{O}(-1)$. We can similarly define $\mathcal{O}(1)$ to be the dual bundle, and $\mathcal{O}(m)$ by taking the $|m|$th tensor product of $\mathcal{O}(1)$ or $\mathcal{O}(-1)$, as appropriate. The transition function for $\mathcal{O}(m)$ is $F = z^m$, and its global sections are homogeneous functions of degree m in (Z^0, Z^1). So we note that

$$\Gamma\big(\mathbb{C}P_1, \mathcal{O}(m)\big) = 0 \quad (m < 0), \qquad \dim\Gamma\big(\mathbb{C}P_1, \mathcal{O}(m)\big) = m+1 \quad (m \geq 0).$$

The second equality can be seen in two ways: first a global section of $\mathcal{O}(m)$ is an entire function $g(z)$ such that $z^{-m}g(z)$ is holomorphic at infinity – that is, g is a polynomial of degree at most m; second, a global homogeneous function of degree m is of the form $\phi_{AB\ldots C}Z^A Z^B \cdots Z^C$ for some symmetric constant 'spinor' $\phi_{AB\ldots C}$.

The same definitions of the line bundles $\mathcal{O}(m)$ can be given over higher-dimensional projective spaces.

6.1.5 Exercises

(1.1) Prove the following variants of Liouville's theorem.

   (i) If M is a compact complex manifold, then every holomorphic function $\phi : M \to \mathbb{C}$ is constant. [Hint: suppose that ϕ is not constant and consider a point where $|\phi|$ achieves its maximum and apply the maximum principle to ϕ as a function of each of the local coordinates to get a contradiction.]

  (ii) If $\phi(z)$ is an entire holomorphic function and ϕ/z^n is bounded for some integer $n \geq 0$, then ϕ is a polynomial of degree at most n.

 (iii) If $\phi(z)$ is a meromorphic function on $\mathbb{C}P_1$ (so its only singularities are a finite collection of poles), then $\phi = s_1/s_2$ for some holomorphic sections s_1, s_2 of $\mathcal{O}(n)$ for some n.

(1.2) Show that if (6.4) holds for some f which is holomorphic in ζ on the whole Riemann sphere, then $F = 0$. [Hint: apply the previous exercise to deduce that f is independent of ζ; then show that Φ is a gradient.]

(1.3) Show that the holomorphic tangent and cotangent bundles of $\mathbb{C}P_1$ are, respectively, $\mathcal{O}(2)$ and $\mathcal{O}(-2)$.

6.2 Lecture 2

6.2.1 Holomorphic bundles

In the case of $\mathbb{C}P_1$, the line bundles $\mathcal{O}(m)$ are in some sense the whole story, since we have the following.

Theorem 1 (Grothendieck) *A rank-k holomorphic vector bundle $E \rightarrow \mathbb{C}P_1$ is isomorphic to a direct sum $\mathcal{O}(m_1) \oplus \cdots \oplus \mathcal{O}(m_k)$ for some integers m_i.*

It is also true that any vector bundle E must have trivial restriction to the z and w coordinate patches. Thus the theorem is equivalent to the statement that the transition matrix

$$F : \mathbb{C} - \{0\} \rightarrow \mathrm{GL}(k, \mathbb{C})$$

can be written in the form

$$F = f^{-1} \begin{pmatrix} z^{m_1} & 0 & \cdots & 0 \\ 0 & z^{m_2} & & 0 \\ & & \ddots & \\ 0 & & & z^{m_k} \end{pmatrix} \tilde{f} \tag{6.5}$$

where $f : U \rightarrow \mathrm{GL}(k, \mathbb{C})$ and $\tilde{f} : \tilde{U} \rightarrow \mathrm{GL}(k, \mathbb{C})$ are holomorphic. Thus Gothendieck's theorem is closely related to Birkhoff's factorization theorem, which states that any smooth map F from the unit circle in the complex plane to $\mathrm{GL}(k, \mathbb{C})$ can be written in this form, where $f, \tilde{f}$ extend holomorphically to the inside and outside of the circle on the Riemann sphere. (It is important to note that the theorem does not hold if 'smooth' is replaced by 'continuous'. Birkhoff's theorem is the link between the twistor view in which solutions to integrable systems are represented by holomorphic vector bundles, and the older approaches in which equations are solved by solving Riemann–Hilbert problem – that is, factorization problems for matrix-valued functions on the circle. A proof is given in Pressley and Segal 1986.)

Example. Put

$$F = \begin{pmatrix} z & 0 \\ t & z^{-1} \end{pmatrix}.$$

For $t = 0$, F is already in the form given by Birkhoff's theorem, with $f = \tilde{f} = 1$, $m_1 = 1$, $m_2 = -1$. But for $t \neq 0$, we have

$$F = \begin{pmatrix} 1 & t^{-1}z \\ 0 & 1 \end{pmatrix} \begin{pmatrix} 0 & -t^{-1} \\ t & z^{-1} \end{pmatrix},$$

so that $m_1 = m_2 = 0$. If we think of F as a patching matrix for a bundle E_t, then $E_0 \simeq \mathcal{O}(1) \oplus \mathcal{O}(-1)$, but E_t is the trivial bundle for $t \neq 0$. This is an example of 'jumping': as t changes through 0, the holomorphic structure of the bundle changes discontinuously, in spite of the fact that the bundles E_t are all the same (and all trivial) from the topological point of view.

6.2.2 Complex space-time

With these preliminaries, we turn to Ward's theorem, which gives a correspondence between anti-self-dual Yang–Mills fields and holomorphic bundles over (open subsets of) of $\mathbb{C}P_3$. It leads to a correspondence between equivariant bundles and solutions to integrable systems that be regarded as symmetry reductions of the anti-self-dual Yang–Mills equations, and thence to more general twistor constructions. To understand its statement and proof, we need some basic facts about conformal geometry in four dimensions (these were covered in more detail by Paul Tod). A much fuller account is given in Ward and Wells (1990).

We shall work almost exclusively with *complex* solutions, the general philosophy being to understand the twistor transforms as entirely holomorphic constructions first, and then to impose reality conditions at a later stage.

We begin, therefore, by representing *complex space-time* as $\mathbb{C}^4$ with coordinates $z, w, \tilde{z}, \tilde{w}$, and metric and volume element

$$\mathrm{d}s^2 = 2(\mathrm{d}z\,\mathrm{d}\tilde{z} - \mathrm{d}w\,\mathrm{d}\tilde{w}) \qquad \nu = \mathrm{d}w \wedge \mathrm{d}\tilde{w} \wedge \mathrm{d}z \wedge \mathrm{d}\tilde{z}.$$

(It should be noted that the apparent signature of the metric is not relevant: in the complex one can change the distribution of plus and minus signs in a diagonal metric by complex coordinates transformation. Conversely the analytic continuations to $\mathbb{C}^4$ of the real metrics on Minkowski space and Euclidean space are the same: one recovers the Euclidean metric by putting $\tilde{z} = \bar{z}$, $\tilde{w} = -\bar{w}$ and the Minkowski metric by putting $\tilde{w} = \bar{w}$, and taking z and $\tilde{z}$ to be real.)

These determine a metric tensor g_{ab} and an alternating tensor ε_{abcd}.

For a 2-form α, we put

$$*\alpha_{ab} = \tfrac{1}{2}\varepsilon_{ab}{}^{cd}\alpha_{cd}\,,$$

and say that α is *self-dual* or *anti self-dual* as $*\alpha = \pm\alpha$. For example

$$dw \wedge dz, \quad d\tilde{w} \wedge d\tilde{z}, \quad dw \wedge d\tilde{w} - dz \wedge d\tilde{z}$$

are self-dual, while

$$dw \wedge d\tilde{z}, \quad d\tilde{w} \wedge dz, \quad dw \wedge d\tilde{w} + dz \wedge d\tilde{z}$$

are anti-self-dual.

Lemma 1 *The duality operation $\alpha \mapsto *\alpha$ does not depend on the scale of the metric.*

A *conformal transformation* $\rho : U \subset \mathbb{C}^4 \to \rho(U)$ is a biholomorphic mapping such $\rho^*(g) \propto g$. If ρ is conformal, then either

$$\rho^*(*\alpha) = *\rho^*(\alpha) \qquad \text{or} \qquad \rho^*(*\alpha) = -*\rho^*(\alpha)$$

for every 2-form α. We shall consider only *proper* transformations, for which the first is true. A central result underlying the twistor constructions, which has been explained from other starting points in other lectures, is that the proper conformal transformations form a group isomorphic to $\mathrm{PGL}(4,\mathbb{C}) \simeq \mathrm{GL}(4,\mathbb{C})/\mathbb{C}^\times$. There are two points to establish here: first, that we do indeed have a group structure (which is awkward because we cannot compose transformations with disjoint domains, while if we insist that $U = \mathbb{C}^4$, then we exclude all but isometries and dilatations). Second, that the transformations can be identified with projective linear transformations.

Both points are dealt with by labelling the points of space-time by 4×4 matrices by putting

$$X = \lambda \begin{pmatrix} 0 & s & -w & \tilde{z} \\ -s & 0 & -z & \tilde{w} \\ w & z & 0 & 1 \\ -\tilde{z} & -\tilde{w} & -1 & 0 \end{pmatrix}$$

for some $\lambda \neq 0$, where $s = z\tilde{z} - w\tilde{w}$. This gives a correspondence between

- points of space-time, and
- skew 4×4 matrices X with $\det X = 0$ and $X_{34} \neq 0$, up to scale.

We then have

Lemma 2 $ds^2 = \sqrt{\det(d\,X)}$.

Note that the right-hand side is in fact quadratic in $d\,X$ because a skew-symmetric matrix has a natural square root (the *Pfaffian*). We have to make a choice for λ: different choices give different metrics in the conformal class.

The isomorphism is then given by $M \in \mathrm{GL}(4, \mathbb{C}) \mapsto \rho$, where

$$\rho(X) = MXM^t. \tag{6.6}$$

Since X is determined only up to scale, M and any nonzero scalar multiple of M give the same transformation, which is why the isomorphism is between the conformal transformations and the projective linear group, rather than $\mathrm{GL}(4, \mathbb{C})$ itself.

One now sees how to address the first point: the action of ρ is not defined by (6.6) on the whole of $\mathbb{C}^4$ because it does not preserve the condition $X_{34} \neq 0$. In fact, this condition looks rather unnatural; if we drop it, and admit matrices that do not satisfy it, then what we have is not space-time itself, but its *conformal compactification*, which we can represent as the quadric hypersurface $\det X = 0$ in the projective space $\mathbb{C}P_5$ (the entries in X are the homogeneous coordinates on $\mathbb{C}P_5$). The original space-time $\mathbb{C}^4$ is the dense open subset $X_{34} \neq 0$, while the extra points are the points of a *light-cone at infinity*. What we have shown is that the group $\mathrm{PGL}(4, \mathbb{C})$ acts on the compactification as a group of conformal transformations, in a similar way to that in which $\mathrm{SL}(2, \mathbb{C})$ acts on the Riemann sphere, which is a compactification of the complex plane. What is a little harder to show is that this action gives all the proper conformal transformations.

6.2.3 The anti-self-dual Yang–Mills equations

We now turn to the anti-self-dual Yang–Mills equations themselves. These are nonlinear differential constraints on a connection D. We shall consider only connections $\mathrm{D} = \mathrm{d} + \Phi$ on the trivial bundle $V \times C^k$, where $V \subset \mathbb{C}^4$ is some open set. We shall thus ignore any topological complications in space-time, and concentrate on the local behaviour of Yang–Mills fields. We shall not, however, assume that the bundle has a fixed trivialization, so we are free to make gauge transformations by $g : V \to \mathrm{GL}(k, \mathbb{C})$, under which

$$\Phi \mapsto g^{-1}\Phi g + g^{-1}\mathrm{d}\,g.$$

The *curvature 2-form*

$$F_{ab} = \partial_a \Phi_b - \partial_b \Phi_a + [\Phi_a, \Phi_b]$$

then transforms homogeneously by $F \mapsto g^{-1}Fg$. Consequently it makes sense to make the gauge-independent definition:

Definition 1 *A connection* D *is a solution to the* anti-self-dual (ASD) *equations if* $F = -*F$.

Equivalently, the operators

$$D_w - \zeta D_{\bar{z}} \quad \text{and} \quad D_z - \zeta D_{\bar{w}}$$

commute for every fixed value of the *spectral parameter* ζ. This is the *Lax pair* or *zero-curvature* formulation of the Yang–Mills equation.

If D is ASD, then $DF = 0$ (the Bianchi identity), and consequently $D*F = 0$; thus, as the terminology implies, an ASD connection is a solution to the Yang–Mills equations. We also have

Proposition 1 *If* $d + \Phi$ *is ASD, then so is* $d + \rho^*\Phi$ *for any proper conformal transformation* ρ.

6.2.4 Exercise

(2.1) Show that $\sqrt{\det(dX)} = ds^2$.

6.3 Lecture 3

6.3.1 Null 2-planes

Ward's theorem is based on the observation that the ASD condition is the integrability condition for D over a special family of 2-planes in $\mathbb{C}^4$.

A 2-plane $Z \subset \mathbb{C}^4$ is *null* if $g(X,Y) = 0$ for every X, Y tangent to Z. We define the *tangent 2-form* ω to such a plane by

$$\omega = \nu(X, Y, \cdot, \cdot)$$

and note that ω is determined by Z up to scale.

Proposition 2 *The tangent 2-form of a null 2-plane is either self-dual or anti-self-dual.*

Proof. The tangent 2-form is characterized (again up to scale) by $\omega(X, \cdot) = 0$ for every X tangent to Z. If Z is null, then $*\omega$ also has this property. Consequently $*\omega = \lambda\omega$ for some λ. But the eigenvalues of $*$ as a linear operator on 2-forms are ± 1. $\square$

Definition 2 *A null 2-plane with a self-dual tangent 2-form is called an* α-*plane.*

The α-planes are given by linear equations of the form

$$Z^0 - \tilde{z}Z^2 - wZ^3 = 0 = Z^1 - \tilde{w}Z^2 - zZ^3 \qquad (6.7)$$

for some constant Z^α, with Z^2, Z^3 not both zero. Since we obtain the same α-plane by rescaling Z^α, we see that the space of α planes are labelled by the points of the projective space $\mathbb{C}P_3$ on which the Z^αs are homogeneous coordinates, excluding the projective line I on which $Z^2 = Z^3 = 0$: this is the twistor space described in Paul Tod's lectures.

6.3.2 Integrability conditions

Proposition 3 *A connection* D *is ASD if and only if the equation* D$s = 0$ *is integrable over every* α-*plane.*

Proof. If we put $\zeta = Z^3/Z^2$, then we can rewrite (6.7) as

$$\tilde{z} + \zeta w = \text{const.}, \qquad \tilde{w} + \zeta z = \text{const.}.$$

So the tangent space to an α-plane is spanned by the vectors

$$L = \partial_w - \zeta\partial_{\tilde{z}}, \qquad M = \partial_z - \zeta\partial_{\tilde{w}},$$

for some constant ζ. Thus the proposition follows by writing the ASD condition as the condition that the operators

$$\mathrm{D}_w - \zeta\mathrm{D}_{\tilde{z}} \qquad \text{and} \qquad \mathrm{D}_z - \zeta\mathrm{D}_{\tilde{w}}$$

should commute. $\square$

The key to Ward's theorem is the observation that the solutions to the equation D$s = 0$ on the α-planes form the fibres of a holomorphic vector bundle over an open subset of the *twistor space* $\mathbb{C}P_3$.

Before we can prove this, we need one other piece of geometry, which will also be familiar from Paul Tod's lectures.

6.3.3 The Klein correspondence

Consider all the α-planes through a given point $x \in \mathbb{C}^4$. These are given by (6.7), but where we now fix $w, z, \tilde{z}, \tilde{w}$ (the coordinates of x) and allow Z^α to vary. We thus have two homogeneous linear equations in the four variables Z^α: they determine a 2-dimensional subspace of $\mathbb{C}^4$ and hence a projective line $\hat{x} \subset \mathbb{C}P_3$.

We thus have a correspondence (the *Klein correspondence*) between points x of $\mathbb{C}^4$ and lines $\hat{x} \subset \mathbb{C}P_3$. Every line that does not intersect I determines a point of $\mathbb{C}^4$ (the lines that do intersect I are the points at infinity in the compactified space-time). The conformal action of $\mathrm{PGL}(4, \mathbb{C})$ on compactified space-time that we looked at above is simply that induced by the natural action of the projective linear group on $\mathbb{C}P_3$.

6.3.4 Ward's theorem

Ward's theorem is an example of a *Penrose transform*, that is, a correspondence between solutions of a differential equation in space-time and holomorphic objects on the corresponding twistor space.

If $U \subset \mathbb{C}^4$ is some subset, then the *twistor space* of U is the set $\mathcal{Z}_U$ of α-planes Z such that $Z \cap U \neq \emptyset$. The twistor space of $\mathbb{C}^4$ itself is $\mathbb{C}P_3 - I$; the twistor space of a single point x is the projective line $\hat{x} \subset \mathbb{C}P_3$. For a general U, we have the incidence condition: if $x \in U$, then $\hat{x} \subset \mathcal{Z}_U$.

With this notation, we have the following.

Theorem 2 (Ward) *Let $U \subset \mathbb{C}^4$ be an open set such that $U \cap Z$ is connected and simply connected for every $Z \in \mathcal{Z}_U$. Then there is a one-to-one correspondence between solutions of the ASD Yang–Mills equation on U with gauge group $\mathrm{GL}(k, \mathbb{C})$ and holomorphic vector bundles $E \to \mathcal{Z}_U$ such that $E|_{\hat{x}}$ is trivial for every $x \in U$.*

We have seen from the general discussion of holomorphic vector bundles that it is possible for $E_{\hat{x}}$ to jump from a trivial bundle to a non-trivial one as x varies. However, it can be shown that its behaviour is semi-continuous in the sense that, if $E_{\hat{x}}$ is trivial, then so is $E|_{\hat{y}}$ for all y some open neighbourhood of x.

Proof. I shall not give the full proof, but simply show how to construct E from D and then indicate in outline how to recover D from E.

Given an ASD connection D with gauge group $\mathrm{GL}(k, \mathbb{C})$, we construct the fibre E_Z of the corresponding bundle $E \to \mathcal{Z}_U$ at $Z \in \mathcal{Z}_U$ by taking E_Z to be the space of solutions to the linear equation $Ds = 0$ on $Z \cap U$. This space is k-dimensional since the linear equation is integrable (by the ASD condition) and because $Z \cap U$ is connected and simply connected, by hypothesis.

It is not immediately obvious that the fibres E_Z fit together to form a holomorphic vector bundle $E \to \mathcal{Z}_U$. To show this, we have to construct holomorphic local trivializations.

Given $x \in U$, we can identify all the spaces E_Z, $Z \in \hat{x}$, with $\mathbb{C}^k$ by evaluating the solutions s of the linear equation at x. If we could find a two-dimensional holomorphic surface $S \subset U$ which intersected each Z in exactly one point, then we could go further and identify every space E_Z with $\mathbb{C}^k$ by evaluating s at $S \cap Z$. This is not in fact possible, but it is almost possible: if we take S itself to be an α-plane through some $x \in U$, then it intersects all other α-planes in exactly one point, with the exception of those with the same (or proportional) tangent 2-form. By making different choices for S, we can identify E_Z with $\mathbb{C}^k$ for Z in various open subsets of U. The transition maps $\mathbb{C}^k \to \mathbb{C}^k$ between the identifications determined by S and S' are given by integrating the equation $Ds = 0$ from $S \cap Z$ to $S' \cap Z$: they are holomorphic functions of the coordinates on $\mathcal{Z}_U$.

Thus $E \to \mathcal{Z}_U$ is a rank-k holomorphic vector bundle. Its restriction to $\hat{x}$ is trivial for any $x \in U$ since $E_Z = \mathbb{C}^k$ for every $Z \in \hat{x}$ by the construction above.

Going in the other direction, suppose that we are given $E \to \mathcal{Z}_U$ with this property; suppose that $\mathcal{Z}_U$ is covered by two open sets V, $\tilde{V}$ on each of which E is trivial and suppose that $\hat{x} \cap V \cap \tilde{V}$ is an annulus for each $x \in U$. Let the transition matrix between these trivializations be $F(\lambda, \mu, \zeta)$, where

$$\lambda = Z^0/Z^2, \qquad \mu = Z^1/Z^2, \qquad \zeta = Z^3/Z^2.$$

Then if the point x has coordinates $w, z, \tilde{w}, \tilde{z}$, the line $\hat{x}$ is given by

$$\lambda = \zeta w + \tilde{z}, \qquad \mu = \zeta z + \tilde{w}. \tag{6.8}$$

We assume that the coordinates have been set up so that $V \cap \hat{x}$ is a disc in the ζ-plane containing $\zeta = 0$, and that $\tilde{V} \cap \hat{x}$ is a disc containing $\zeta = \infty$, and that $\hat{x} \cap V \cap \tilde{V}$ is an annular neighbourhood of the unit circle in the ζ-plane.

Since $E|_{\hat{x}}$ is trivial, we must have a Birkhoff factorization of the form

$$F(\zeta w + \tilde{z}, \zeta z + \tilde{w}, \zeta) = f^{-1}\tilde{f}, \tag{6.9}$$

where f and $\tilde{f}$ are functions of the five variables $w, z, \tilde{w}, \tilde{z}, \zeta$ with values in $\mathrm{GL}(k, \mathbb{C})$, with f holomorphic at $\zeta = 0$ and $\tilde{f}$ holomorphic at $\zeta = \infty$. A crucial point, however, is that f and $\tilde{f}$ will not in general be expressible as functions of λ, μ and ζ (this happens only when E itself is trivial). Now

$$\partial_w F - \zeta \partial_{\tilde{z}} F = 0.$$

Hence

$$\partial_w f f^{-1} - \zeta \partial_{\tilde{z}} f f^{-1} = \partial_w \tilde{f} \tilde{f}^{-1} - \zeta \partial_{\tilde{z}} \tilde{f} \tilde{f}^{-1}.$$

Both sides, therefore, must be equal to a global rational function of ζ with a simple pole at infinity (see Exercise 1.1). Therefore

$$\partial_w f f^{-1} - \zeta \partial_{\tilde{z}} f f^{-1} = -\Phi_w + \zeta \Phi_{\tilde{z}}$$

where Φ_w and $\Phi_{\tilde{z}}$ are matrix-valued functions of the space-time coordinates (but not of ζ). By repeating this argument for the operator $\partial_z - \zeta \partial_{\tilde{w}}$, we get that f and $\tilde{f}$ are solutions of a linear system of the form

$$\begin{aligned} (\partial_w + \Phi_w)f - \zeta(\partial_{\tilde{z}} + \Phi_{\tilde{z}})f &= 0 \\ (\partial_z + \Phi_z)f - \zeta(\partial_{\tilde{w}} + \Phi_{\tilde{w}})f &= 0, \end{aligned}$$

the integrability condition for which (for all constant ζ) is precisely that $d + \Phi$ should be an ASD connection. $\square$

The transition matrix can be chosen freely; thus the result reduces the solution of the ASD Yang–Mills equation to the solution of a *Riemann–Hilbert problem* – the factorization (6.9). In the proof, we made assumptions about the covering V, $\tilde{V}$; these can, in fact, be justified. Alternatively, one can work with a more general cover, in which case the factorization problem is of the form $F_{ij} = f_i^{-1} f_j$.

6.3.5 Exercises

(3.1) Show that the factorization (6.9) is unique up to multiplication of f and $\tilde{f}$ on the left by a nonsingular matrix g depending on the space-time coordinates, but not ζ. Show that different choices of factorization give gauge-equivalent connections.

(3.2) Show that if, in the proof of Ward's theorem, f and $\tilde{f}$ can be expressed as functions of λ, μ and ζ, then D is gauge-equivalent to $\Phi = 0$.

(3.2) In the proof of Ward's theorem, show that the if D is constructed from E by Birkhoff factorization, and that if E' is constructed from D, then $E \simeq E'$.

(3.3) Suppose that

$$F = \begin{pmatrix} 1 & h \\ 0 & 1 \end{pmatrix}$$

where $h = h(\lambda, \mu, \zeta)$. Find Φ in terms of the electromagnetic field generated by h.

(3.4) Show that the natural action of $\mathrm{PGL}(4, \mathbb{C})$ on $\mathbb{C}P_3$ induces an action on the space of lines in $\mathbb{C}P_3$ and hence an action on space-time. Show that this coincides with the conformal action described in Lecture 2. [Hint: the notation of Paul Tod's lectures helps here.]

6.4 Lecture 4

6.4.1 Equivariant holomorphic bundles

In Lionel Mason's lectures, it has been shown how various integrable systems of nonlinear differential equations arise as reductions of the ASD Yang–Mills equations by subgroups of the conformal group. We shall now introduce Penrose transforms for these equations by imposing symmetry on Ward's construction.

A subgroup of the conformal group is the same thing as a subgroup of the projective linear group $\mathrm{PGL}(4, \mathbb{C})$, which acts transitively on the projective space $\mathbb{C}P_3$, by the linear action of $\mathrm{GL}(4, \mathbb{C})$ on the homogeneous coordinates Z^α. The generators are the holomorphic vector fields of the form

$$Y_A = A^\alpha{}_\beta Z^\beta \frac{\partial}{\partial Z^\alpha},$$

for some constant 4×4 matrix A (note that Y is well defined on $\mathbb{C}P_3$ because the expression on the right-hand side is homogeneous in Z of degree zero). Thus to solve some reduced form of the ASD Yang–Mills equation, we are led to consider holomorphic bundles E which are *equivariant* under under this action on $\mathbb{C}P_3$.

In general, of course, an element of the conformal group does not give a well-defined mapping $\mathcal{Z}_U \to \mathcal{Z}_U$ since it may move an α-plane that intersects U to one that does not. For this reason, we concentrate on the equivariance condition at the Lie algebra level. Before making this precise, we need some definitions and some general results about equivariant bundles.

Definition 3 *Let $E \to M$ be a holomorphic vector bundle and let $\rho : M \to M$ be a biholomorphic transformation. We say that E is equivariant under ρ if ρ lifts to a biholomorphic map $\hat{\rho} : E \to E$ such that $\hat{\rho}$ maps the fibre E_m linearly onto the fibre $E_{\rho(m)}$ for each $m \in M$.*

Thus an equivariant bundle is one for which $\rho^* E \simeq E$ (the pull-back bundle $\rho^* E$ is the bundle with fibres $(\rho^* E)_m = E_{\rho(m)}$). In practical terms, if E has transition matrices F_{ij} relative to some cover U_i, then $\rho^* E$ has transition matrices $F_{ij} \circ \rho$ relative to the cover $\rho^{-1} U_i$.

As usual, one should be careful about taking over intuitive ideas from the behaviour of smooth bundles: if ρ is close to the identity, then every smooth bundle is equivariant along ρ (in the sense of the above definition, with holomorphic maps replaced by smooth ones); but this is certainly not true in general for holomorphic bundles.

At the Lie algebra level, equivariance is characterized by the existence of a lift of a Lie algebra of vector fields on M to vector fields on E. More conveniently, we can think in terms of the existence of 'Lie derivative operators'.

Definition 4 *Let $E \to M$ be a holomorphic vector bundle over a complex manifold M and let Y be a holomorphic vector field on M. We say that E is equivariant along Y if there exists a first-order differential operator $\mathcal{L}_Y$ on the local sections of E which takes the form*

$$\mathcal{L}_Y = Y + \theta_Y$$

for some holomorphic matrix-valued function θ_Y in a local trivialization.

We note that if $\mathcal{L}_Y$ exists, then θ_Y transforms by

$$\theta_Y \mapsto g^{-1} \theta_Y g + g^{-1} Y(g) \tag{6.10}$$

under transformations of the local trivialization. If we introduce local coordinates z^a on M and linear coordinates w^i on the fibres of E, then

$$\hat{Y} = Y^a \frac{\partial}{\partial z^a} - \theta^i{}_j w^j \frac{\partial}{\partial w^i}$$

is a vector field on the total space of E, where the Y^as are the components of Y and the $\theta^i{}_j$s are the entries in the matrix θ_Y. If we can integrate Y to construct a family of biholomorphic transformations ρ, then by integrating $\hat{Y}$, we obtained the lifted transformations $\hat{\rho}$. Thus Definition (4) is indeed an infinitesimal form of Definition (3).

If E is equivariant and is given by a transition matrix F between local trivializations over the V and $\tilde{V}$ (where $M = V \cup \tilde{V}$), then

$$Y(F) = -\theta_Y F + F\tilde{\theta}_Y,$$

where $\mathcal{L}_Y = Y + \theta_Y$ in V and $\mathcal{L}_Y = Y + \tilde{\theta}_Y$ in $\tilde{V}$. Conversely, if we can write $Y(F)$ in this form, with θ_Y holomorphic in V and $\tilde{\theta}_Y$ holomorphic in $\tilde{V}$, then E is equivariant.

If $Y(m) \neq 0$ for some point m, then we can solve the equation

$$Y(g) + \theta_Y g = 0$$

for $g : V \to \mathrm{GL}(k, \mathbb{C})$ in some neighbourhood V of Y to obtain a local trivialization of E such that $\theta_Y = 0$. Any other local trivialization with this property is then given by transforming the local frames by some invertible matrix-valued function F such that $Y(F) = 0$. It follows that if Y has no zeros, then we can choose local trivializations of E such that the transition functions are constant along Y. If it is also true that the space of integral curves of Y is Hausdorff manifold M', and that the curves are simply connected, then a Y-equivariant bundle is the same thing as the pull-back of a holomorphic vector bundle $E' \to M'$ by the projection map $M \to M'$.

Example. The Bogomolny equations Consider the *weighted projective space* given by taking $\mathbb{C}^3 - \{W^2 = W^3 = 0\}$ and making the identification

$$(W^1, W^2, W^3) \sim (\lambda^2 W^1, \lambda W^2, \lambda W^3) \qquad \lambda \neq 0 \in \mathbb{C}.$$

This is a two-dimensional complex manifold, which is fibred over $\mathbb{C}P_1$ by the projection $(W^1, W^2, W^3) \mapsto (W^2, W^3)$. (It is in fact the total space of the tangent bundle $T\mathbb{C}P_1$). We also have a projection

$$\mathbb{C}P_3 - I \to T\mathbb{C}P_1 \;:\; (Z^0, Z^1, Z^2, Z^3) \mapsto (W^1, W^2, W^3)$$
$$= (Z^3 Z^1 + Z^2 Z^0, Z^2, Z^3).$$

The fibres of this are tangent to the vector field

$$Y = Z^3 \frac{\partial}{\partial Z^0} - Z^2 \frac{\partial}{\partial Z^1} = \zeta \frac{\partial}{\partial \lambda} - \frac{\partial}{\partial \mu},$$

which has no zeros on $\mathbb{C}P_3 - I$. We thus have a correspondence between holomorphic vector bundles $E' \to T\mathbb{C}P_1$ and Y-equivariant holomorphic vector bundles $E \to \mathbb{C}P_3 - I$.

Now Y is a generator of a conformal action on space-time. In fact, we see from (6.8) that it generates the translation

$$\frac{\partial}{\partial w} - \frac{\partial}{\partial \tilde{w}}.$$

Thus we expect a correspondence between (i) solutions of the ASD Yang–Mill equations which are invariant under translation by $\partial_w - \partial_{\tilde{w}}$ and (ii) holomorphic vector bundles over TCP_1. The corresponding symmetry reductions of the ASD Yang–Mills equations are the *Bogomolny equations*

$$F = *\mathrm{D}\phi\,,$$

where D is a connection on a vector bundle B over some region in three-dimensional Euclidean space (complexified), ϕ (the 'Higgs field') is a section $\mathrm{adj}(B)$, and the $*$ is the three-dimensional duality operator.

This is in fact true, and has been exploited by Hitchin (1982) in the construction of monopoles.

6.4.2 Normal forms

The situation is more interesting if Y has zeros (so that its flow has fixed points). If $Y(m) = 0$, then $g^{-1}Y(g) = 0$ at m and so $\theta_Y(m)$ transforms by conjugation under change of local trivialization. Thus if $\theta_Y(m) \neq 0$ in some local trivialization, then $\theta_Y(m) \neq 0$ in every local trivialization. In fact the flow of Y induces a one (complex) parameter family of linear transformations of the fibre E_m over the fixed point, generated by $\theta_Y(m)$. The best that we can do in this case is to reduce θ_Y to normal form in some neighbourhood of m.

The full range of possible behaviours is very complicated; we shall just look at some simple examples, which give the flavour of the sort of phenomena one can expect, and indicate the connection with the classification of singular points of ordinary linear differential equations. In the following, we suppose that z and w are two coordinates on a complex manifold.

Logarithmic fixed points. Suppose that $\dim M = 1$ and that $Y = z\partial_z$. This has a zero at the point $z = 0$. If $\theta = \theta_Y(0)$ has distinct eigenvalues, no two of which differ by an integer, then we can choose a local trivialization in which θ is a diagonal matrix m, and we can find a fundamental matrix solution y to the ordinary differential equation

$$z\frac{\mathrm{d}y}{\mathrm{d}z} + \theta_Y y = 0$$

of the form $y = h(z)z^{-m}$, where h is a power series in z. That is,

$$h^{-1}Y(h) + h^{-1}\theta_Y h = m\,.$$

Thus we can use h to reduce θ_Y to m everywhere (in a neighbourhood of

$z = 0$). Because the solution y is not single-valued in general, we cannot make $\theta_Y = 0$.

Irregular singular points. If we take instead $Y = z^2 \partial_z$, then the problem of finding a good local normal form is more involved. We start in this case with the linear ordinary differential equation

$$z^2 \frac{dy}{dz} + \theta_Y y = 0, \quad \text{or} \quad \frac{dy}{dz} = Ay$$

where $A = \theta_Y/z^2 = c_{-2}/z^2 + c_{-1}/z + c_0 + \cdots$ is a matrix-valued function of z with a double pole at the origin (the c_i are the coefficients in its Laurent expansion). Provided that the eigenvalues of c_{-2} are distinct, we can can choose a local trivialization in which c_{-2} is diagonal, and we can find a *formal* solution of the form

$$y = h \exp\left(-c_{-2}/z + m \log z\right)$$

where $h = 1 + h_1 z + h_2 z^2 + \cdots$ is a formal power series with matrix coefficients and m is a constant diagonal matrix. If this power series converges, then

$$h^{-1} Y(h) + h^{-1} \theta_Y h = -mz - c_{-2}$$

and so we can reduce the Lie derivative operator to diagonal form, with θ_Y depending linearly on z. The constant term gives the action of $\mathcal{L}_Y$ on the fibre above the fixed point. In general, however, the series does not converge. By truncating it, however, at some power of z, we can achieve

$$\theta_Y = -mz - c_{-2} + O(z^p),$$

for p arbitrarily large.

Dressing. Another type of behaviour will be important in the KdV example below. Suppose that M is 2-dimensional and that $Y = z^2 \partial_w$. This vanishes on the z-axis. Let us suppose in this case that

$$\theta_Y = N + z^2 U \qquad N = \begin{pmatrix} 0 & 1 \\ z & 0 \end{pmatrix},$$

where $\operatorname{tr} U = 0$ and U is upper triangular at $z = 0$. In this case, the linear action on the fibre above the fixed point has the nilpotent generator

$$\begin{pmatrix} 0 & 1 \\ 0 & 0 \end{pmatrix}.$$

The question that arises here is: can we remove U by changing the local trivialization? The answer is that we can, but again only formally, by the 'dressing procedure' described by Drinfeld and Sokolov (1985). The idea is to begin by noting that U can be written as a convergent power series

$$U = \sum_0^\infty \begin{pmatrix} h_i + \alpha_i & 0 \\ 0 & h_i - \alpha_i \end{pmatrix} N^i$$

where the coefficients h_i, α_i depend on w alone. We can reduce the coefficients α_i to zero by an iterative procedure. Suppose that i is the least index such that $\alpha_i \neq 0$ and put

$$g = 1 + \begin{pmatrix} \alpha_i & 0 \\ 0 & 0 \end{pmatrix} N^{i-1}.$$

Then, since $z^2 = N^4$ and since

$$\left[\begin{pmatrix} \alpha_i & 0 \\ 0 & 0 \end{pmatrix}, \begin{pmatrix} 0 & 1 \\ z & 0 \end{pmatrix} \right] = \begin{pmatrix} \alpha_i & 0 \\ 0 & -\alpha_i \end{pmatrix} \begin{pmatrix} 0 & 1 \\ z & 0 \end{pmatrix}$$

we have that $g^{-1}Y(g) + g^{-1}\theta_Y g$ has an expansion with i increased by one. By this algebraic procedure, therefore, we can arrange that

$$\theta_Y = N + z^2 \sum_0^{2p} h_i N^i + O(z^{p+2})$$

for any positive p. A further iteration, where now we take at each stage g to be of the form $g = 1 + g_i(w)N^i$ for scalar $g_i(w)$ reduces the h_is to zero up to any finite i by solving a differential equation for g_i as a function of w at each stage. Thus for any positive p, we can choose the frame in a neighbourhood of the origin so that

$$\mathcal{L}_Y = z^2 \partial_w + N + O(z^p).$$

Formally, we can take the procedure to the limit; but the relevant power series do not converge in general.

Finally, we remark that we have only considered here equivariance along a single vector field. More generally, E is said to be equivariant under a Lie algebra $\mathfrak{h}$ of holomorphic vector fields if it is equivariant along each $Y \in \mathfrak{h}$, and if $Y \mapsto \mathcal{L}_Y$ is a representation of $\mathfrak{h}$ by differential operators.

6.4.3 Exercises

(4.1) Prove that every holomorphic line bundle over the Riemann sphere is equivariant under the action of $SL(2, \mathbb{C})$.

(4.2) Show that if E is a smooth bundle on a smooth manifold M, then E is equivariant along any smooth vector field on M. [Hint: piece together local Lie derivative operators by using a partition of unity.]

6.5 Lecture 5

6.5.1 The KdV equation

In Lionel Mason's lectures, he shows that the KdV equation is a symmetry reduction of the ASD Yang–Mills equation by two orthogonal translations, one null and the other non-null (a result due to Mason and Sparling 1989). In our notation, these are the two commuting vector fields

$$X' = \partial_{\tilde{z}}, \qquad Y' = \partial_w - \partial_{\tilde{w}}.$$

We have already seen that the second corresponds to the vector field

$$Y = Z^3 \frac{\partial}{\partial Z^0} - Z^2 \frac{\partial}{\partial Z^1} = \zeta \frac{\partial}{\partial \lambda} - \frac{\partial}{\partial \mu},$$

on $\mathbb{C}P_3$. The first corresponds to $X = Z^2 \partial / \partial Z^0$.

In the example of the Bogomolny equation above that we saw that we can take a quotient along Y to reduce a Y-equivariant bundle to one over $T\mathbb{C}P_1$. The projection of X onto this space is ∂_γ, which we also denote by X, in the coordinates

$$\zeta = Z^3/Z^2, \qquad \gamma = (Z^3 Z^1 + Z^2 Z^0)/(Z^2)^2 = \mu\zeta + \lambda.$$

These do not, of course, cover the fibre over the point $\zeta = \infty$: there we must use another coordinate system, for example

$$\tilde{\zeta} = Z^3/Z^2 = 1/\zeta, \qquad \tilde{\gamma} = (Z^3 Z^1 + Z^2 Z^0)/(Z^3)^2 = \gamma/\zeta^2.$$

In these, $X = \tilde{\zeta}^2 \partial_{\tilde{\gamma}}$, so that X vanishes at $\zeta = \infty$. Thus we cannot take a further quotient (which is fortunate, because otherwise the whole construction would be trivial).

Each point in space-time determines a line in $\mathbb{C}P_3$ and hence an embedded copy of $\mathbb{C}P_1$ in $T\mathbb{C}P_1$. From (6.8), this is given in terms of the coordinates $w, z, \tilde{w}, \tilde{z}$ by

$$\gamma = z\zeta^2 + (w + \tilde{w})\zeta + \tilde{z}.$$

That is, it is a section of the bundle $T\mathbb{C}P_1 \to \mathbb{C}P_1$

The Yang–Mills theory suggests that there should be a correspondence between

- solutions to the KdV equation which are holomorphic in $t = z$ and $x = w + \tilde{w}$ in some region of the t, x-plane and
- holomorphic vector bundles over some corresponding region in $T\mathbb{C}P_1$ which are equivariant along X and trivial on the copies of $\mathbb{C}P_1$.

Not every such bundle will do: it is important that the Lie derivative operator $\mathcal{L}_X$ should have the right normal form at the zeros of X.

We cannot here go into the detailed argument that leads to the correct normal form. Essentially there are two possibilities (apart from the trivial one): that θ_X should be either be semi-simple or nilpotent at $\tilde{\zeta} = 0$. The choice is determined by the form of the Higgs field corresponding to X': for the KdV equation it is nilpotent; for the nonlinear Schrödinger equation, it is semi-simple. However the story is not quite that simple (a lot is hidden in the qualification 'essentially').

What we shall do instead of pursuing the full details of the reduction of the ASD Yang–Mills equation is to show directly that the bundles with a particular normal form give rise to solutions of the KdV equation.

6.5.2 Solutions to the KdV equation

Suppose that E is a holomorphic $SL(2, \mathbb{C})$-vector bundle E over a neighbourhood of the zero section in $T\mathbb{C}P_1$ satisfying the triviality condition on sections of $T\mathbb{C}P_1$ and and equivariant along

$$X = \partial_\gamma = \tilde{\zeta}^2 \partial_{\tilde{\gamma}} \,.$$

The KdV bundles are distinguished by the form of $\mathcal{L}_X$ in a neighbourhood of the fibre at infinity ($\tilde{\zeta} = 0$), where X vanishes. They satisfy the condition that there should exist a frame in a neighbourhood of $\zeta = \infty$ in which

$$\mathcal{L}_X = \partial_\gamma - \zeta^{-1}\Lambda + \zeta^{-2}U \,; \qquad \Lambda = \begin{pmatrix} 0 & \zeta \\ 1 & 0 \end{pmatrix},$$

where U is upper triangular at $\zeta = \infty$. By the iterative procedure described above, it is then possible to change the frame so that $U = O(\zeta^{-p})$ as $\zeta \to \infty$ for arbitrary $p > 0$.

If we use this together with an invariant frame over the complement of

$\zeta = \infty$ to construct local trivializations of $E \to \mathcal{O}(2)$, then the transition matrix $F(\zeta, \gamma)$ satisfies

$$F^{-1}\frac{\partial F}{\partial \gamma} = -\zeta^{-1}\Lambda + O(\zeta^{-p-2})$$

as $\zeta \to \infty$.

The solution itself is recovered from F by substituting $\gamma = x\zeta + t\zeta^2$, and making a Birkhoff factorization $F = f^{-1}\tilde{f}$ in ζ, with $\tilde{f} = 1$ at infinity (this fixes the gauge). If this can be found, then

$$\partial_x f f^{-1} = \partial_x \tilde{f}\tilde{f}^{-1} - \tilde{f}\Lambda\tilde{f}^{-1} + O(\zeta^{-p-1})$$
$$\zeta\partial_x f f^{-1} - \partial_t f f^{-1} = \zeta\partial_t \tilde{f}\tilde{f}^{-1} - \partial_x \tilde{f}\tilde{f}^{-1},$$

since $\zeta\partial_x F = \partial_t F = \zeta^2\partial_\gamma F$. Both sides of the two equations must be global rational functions of ζ. By examining their behaviour at $\zeta = \infty$, we conclude that the two sides of the two equations are respectively of the forms $-\Lambda + A$ and B, where A and B depend only on x, t, and A is upper triangular. Therefore

$$\partial_x f + Af - \Lambda f = 0, \qquad \partial_t f + Bf - \zeta\partial_x f = 0,$$

and the two operators on the left-hand sides commute. The integrability of this linear system implies that

$$A = \begin{pmatrix} -q & r \\ 0 & q \end{pmatrix}, \qquad B = \begin{pmatrix} * & * \\ -q_x & * \end{pmatrix}.$$

Now $\tilde{f}$ also satisfies the second linear equation, while instead of the first, it satisfies

$$\partial_x \tilde{f} = \tilde{f}\Lambda - \Lambda\tilde{f} + A\tilde{f} + O(\zeta^{-p-1}) \tag{6.11}$$

But

$$\tilde{f} = 1 + \zeta^{-1}\begin{pmatrix} \sigma & \rho \\ \tau & -\sigma \end{pmatrix} + O(\zeta^{-2})$$

as $\zeta \to \infty$, for some σ, ρ, τ depending on x, t. By substituting this into (6.11), we deduce that

$$r = -2\sigma, \qquad \tau = -q, \qquad \tau_x + 2\sigma = \tau q,$$

and hence that $r = q_x - q^2$. The commutation relation now gives that $v = 2q_x$ satisfies the KdV equation $4v_t - v_{xxx} - 6vv_x = 0$. In fact every local analytic solution of the KdV solution arises in this way.

We can generate solutions by taking

$$F(\zeta, \gamma) = g(\zeta)e^{-\gamma\Lambda/\zeta}$$

where g is a holomorphic $SL(2, \mathbb{C})$-valued function on an annulus in the ζ plane. This is the case in which the iteration converges, since

$$\frac{\partial F}{\partial \gamma} = -FΛ/\zeta.$$

For example, when g is rational, we obtain a multi-soliton solution.

Given one patching matrix $F(\zeta, \gamma)$ satisfying the equivariance condition, we can obtain others by replacing F by

$$\hat{F}(\zeta, \gamma) = F\big(\zeta, \gamma + h(\zeta)\big),$$

where $h(\zeta)$ is holomorphic in ζ is a neighbourhood of $\zeta = 0$. If we think of the coefficients in the Taylor expansion of h as the 'times', then the result is the *KdV hierarchy*.

6.5.3 The isomonodromy problem

Finally, we turn to the twistor theory of the Painlevé equations and the general isomonodromy deformation equations. The aim is to gain some geometric insight into why these lie at the heart of the theory of integrable systems.

6.5.4 Fuchsian equations

Suppose that we have an ordinary equation of the form

$$\frac{\mathrm{d}y}{\mathrm{d}\zeta} = Ay \tag{6.12}$$

where y takes values in the $k \times k$ matrices and $A(\zeta)$ is a matrix-valued function of ζ with rational coefficients. In the *Fuchsian* case, A has only simple poles, so that

$$A = \sum_{1}^{N} \frac{A_\alpha}{\zeta - a_\alpha} \tag{6.13}$$

where the residues A_α are constant matrices.

There are two simple transformations that preserve the Fuchsian form: first, *gauge transformations* (for constant g)

$$A \mapsto g^{-1}Ag, \qquad y \mapsto g^{-1}y;$$

and second Möbius transformations of the coordinate

$$\zeta \mapsto \frac{p\zeta + q}{r\zeta + s}.$$

The matrix A has poles at the points a_α; if

$$A_0 = -A_1 - A_2 - \cdots - A_N$$

is nonzero, then it has a further simple pole at $\zeta = \infty$ with residue A_0. We see this by making the transformation $\zeta \mapsto 1/\zeta$. Thus in general there are $N + 1$ poles.

We also note that if y is invertible at one point, then it is invertible everywhere (except at the poles), and that any other solution is of the form yM for some constant matrix M.

The elementary theory of ordinary differential equations tells us that for any point ζ_0 (other than a pole) there is a unique local holomorphic solution y such that $y(\zeta_0) = 1$. However it is not single-valued in the large. If we analytically continue y around a closed path γ in the complement of the poles, with endpoints at ζ_0, then it it does not return to its original value at ζ_0, but to M_γ (an element of $\mathrm{GL}(k, \mathbb{C})$). This depends only on the homotopy class of γ. The map

$$\pi_1(\mathbb{C} - \{a_1, \ldots, a_N\}) \to \mathrm{GL}(N, \mathbb{C}) : \gamma \mapsto M_\gamma$$

is a representation of the fundamental group π_1, called the *monodromy representation*. The *isomonodromy problem* is to find all the Fuchsian equations with the same monodromy as the given one. Since we can easily change the representation to a conjugate one,

$$\gamma \mapsto g^{-1}M_\gamma g, \qquad g \text{ constant},$$

by changing ζ_0 or by making a gauge transformation, we understand 'same' in this context to mean 'same up to conjugacy'. (The related *Riemann–Hilbert problem* is to find a Fuchsian equation with given monodromy.)

6.5.5 Exercises

(5.1) Show that the vector fields Y and X in the KdV example are global on $\mathbb{C}P_3$. [Hint: find expressions for them in homogeneous coordinates.]

(5.2) Show that

$$\exp(-\gamma\Lambda/\zeta) = \begin{pmatrix} \cosh(\gamma/\zeta^{1/2}) & -\zeta^{1/2}\sinh(\gamma/\zeta^{1/2}) \\ -\zeta^{-1/2}\sinh(\gamma/\zeta^{1/2}) & \cosh(\gamma/\zeta^{1/2}) \end{pmatrix}.$$

(5.3) Let $\alpha, \beta \in \mathbb{C}$ with $|\alpha|, |\beta| < 1$ and let $C : \mathbb{C} \to \mathrm{SL}(2, \mathbb{C})$ be entire. Show that if

$$F(\zeta) = \frac{1}{\zeta - \alpha} \begin{pmatrix} \zeta - \beta & 0 \\ 0 & \zeta - \alpha \end{pmatrix} C(\zeta)^{-1}$$

and

$$\tilde{f}(\zeta) = 1 - \frac{1}{M(\zeta - \alpha)} C(\beta) \begin{pmatrix} 1 & 0 \\ 0 & 0 \end{pmatrix} C(\alpha)^{-1}$$

then (i) $\tilde{f}$ is holomorphic and invertible for all ζ outside of the unit disc (including $\zeta = \infty$), (ii) $\det \tilde{f} = (\zeta - \beta)/(\zeta - \alpha)$, and (iii) $f = \tilde{f} F^{-1}$ is holomorphic and invertible inside the unit disc. Hence find explicitly the solution to the KdV equation generated by

$$F(\gamma, \zeta) = \frac{1}{\zeta - \alpha} \begin{pmatrix} \zeta - \beta & 0 \\ 0 & \zeta - \alpha \end{pmatrix} \exp(-\gamma \Lambda/\zeta) .$$

[Hints: use Exercise (5.2); use the fact that $\tilde{f}$ satisfies the second linear equation to find q_x.]

6.6 Lecture 6

6.6.1 Fuchsian equations from equivariant bundles

The twistor solution to the isomonodromy problem comes from a correspondence between Fuchsian equations and equivariant holomorphic bundles over (subsets) of $\mathbb{C}P_N$.†

Let $\mathcal{Z} \subset \mathbb{C}P_N$ be a neighbourhood of a projective line $X_0 \subset \mathbb{C}P_N$ and let $E \to \mathcal{Z}$ be a rank-k holomorphic vector bundle such that

- $E|_{X_0}$ is trivial;
- E is equivariant under the Lie algebra of the diagonal subgroup of $\mathrm{PGL}(N + 1, \mathbb{C})$.

The diagonal subgroup is generated by any N of the $N + 1$ commuting vector fields

$$Y_0 = Z^0 \frac{\partial}{\partial Z^0}, \quad Y_1 = Z^1 \frac{\partial}{\partial Z^1}, \quad \ldots, \quad Y_N = Z^N \frac{\partial}{\partial Z^N}$$

(their sum is everywhere zero). However, rather than discard one generator, it is simpler to treat all the vector fields on the same footing. So

---

† The connections between the Painlevé equations and twistor methods were first explored by Hitchin (1995). Various reductions of the ASD Yang-Mills equations to the Painlevé equations are catalogued in Ablowitz and Clarkson (1991).

if we denote the corresponding Lie derivatives by $\mathcal{L}_\alpha = \mathcal{L}_{Y_\alpha} = Y_\alpha + \theta_\alpha$ (in a local trivialization), then for consistency we shall require

$$\sum_0^N \mathcal{L}_\alpha = \sum_0^N \theta_\alpha = 0,$$

as well as the commutation relations $[\mathcal{L}_\alpha, \mathcal{L}_\beta] = 0$.

Except on the coordinate hyperplanes $Z^\alpha = 0$, the vector fields span the tangent space to $\mathbb{C}P_N$. We denote the union of these by Σ and define a connection ∇ on the restriction of E to $\mathcal{Z} - \Sigma$ by taking

$$\nabla_T = \sum t^\alpha \mathcal{L}_\alpha, \quad \text{where } T = \sum t^\alpha Y_\alpha.$$

In fact if $T = T^\alpha \partial/\partial Z^\alpha$, then $t_\alpha = T^\alpha/Z^\alpha$. By the consistency condition, this is independent of the way in which T is written as a combination of the Y_αs; and, by the commutation relation, ∇ is flat.

Now suppose that X is a line near X_0. Then $E|_X$ is also trivial, and so we can write can write $E|_X = X \times \mathbb{C}^k$. It is claimed that (i) the restriction of ∇ to X is a Fuchsian differential operator of the form

$$\mathrm{d} - A\,\mathrm{d}\zeta$$

where A is as above and ζ is a stereographic coordinate on X; and (ii), as X changes, the mononodromy of the Fuchsian equation

$$\frac{\mathrm{d}y}{\mathrm{d}\zeta} = Ay$$

is preserved.

To prove the first statement, we write the equation of X in the form

$$Z^\alpha = B^\alpha - \zeta C^\alpha \qquad \alpha = 0, 1, \ldots, N$$

for constants B^α and C^α. The tangent vector $T = \mathrm{d}/\mathrm{d}\zeta$ can be written

$$T = -\sum_\alpha C^\alpha \frac{\partial}{\partial Z^\alpha} = \sum_\alpha \frac{C^\alpha}{B^\alpha - \zeta C^\alpha} Y_\alpha.$$

Thus

$$\nabla_T = \frac{\mathrm{d}}{\mathrm{d}\zeta} - \sum_\alpha \frac{\theta_\alpha}{\zeta - a_\alpha}$$

where $a_\alpha = B^\alpha/C^\alpha$. If we now work in the global trivialization of $E|_X$, then the sum on the right-hand side a global rational function on the Riemann sphere with simple poles at the points $\zeta = a_\alpha$. So we have an operator of the required form (except that the poles $\zeta = a_\alpha$ are in general position: to move a_0 to $\zeta = \infty$, we have to make a Möbius transformation).

To prove the second statement, we note that the monodromy of the Fuchsian equation is simply the holonomy of ∇ around the singularities on the coordinate hyperplanes; and this is independent of the choice of X.

Thus the construction gives us, from an equivariant vector bundle, a family of Fuchsian equations with the same monodromy. The poles of the equation are the intersection points of X with the coordinate hyperplanes. These are $N + 1$ points in general position. Provided $N > 2$, we cannot in general map the poles for one choice of X to those of another by a Möbius transformation (since the cross-ratio of any four poles is preserved by a Möbius transformation). Thus the equations on different lines are genuinely distinct.

6.6.2 Explicit calculation

Suppose that E is given by a patching matrix $F(Z^\alpha)$, between trivializations over a two-set open cover $U, \tilde{U}$ of $\mathcal{Z}$. Since F is a function on $\mathbb{C}P_N$, it is homogeneous of degree zero in the coordinates Z^α. As in the Ward construction, we make the simplifying assumption that $X \cap U$ is a neighbourhood of $\zeta = 0$, $X \cap \tilde{U}$ is a neighbourhood of $\zeta = \infty$ and $X \cap U \cap \tilde{U}$ is an annular neighbourhood of the unit circle in the ζ-plane.

There is no loss of generality in restricting to the case $C^\alpha = 1$, $B^\alpha = a_\alpha$. Since $E|_X$ is trivial, we can find $f, \tilde{f}$ such that

$$F(a_\alpha - \zeta) = f^{-1}\tilde{f} \tag{6.14}$$

where $f : X \cap U \to \mathrm{GL}(k, \mathbb{C})$ and $\tilde{f} : X \cap \tilde{U} \to \mathrm{GL}(k, \mathbb{C})$ are holomorphic. Then as X varies, f and $\tilde{f}$ become holomorphic functions of a_α and ζ.

By the equivariance condition

$$Y_\alpha F = -\theta_\alpha F + F\tilde{\theta}_\alpha$$

where $Y_\alpha + \theta_\alpha$ and $Y_\alpha + \tilde{\theta}_\alpha$ are the Lie derivatives in the two trivializations of E. Also, with $Z^\alpha = a_\alpha - \zeta$, we have

$$\partial_\zeta F = -\sum_0^N \frac{\partial F}{\partial Z^\alpha} = \sum_0^N \frac{\theta_\alpha F - F\tilde{\theta}_\alpha}{a_\alpha - \zeta}$$

$$\partial_{a_\alpha} F = \frac{\partial F}{\partial Z^\alpha} = \frac{-\theta_\alpha F + F\tilde{\theta}_\alpha}{a_\alpha - \zeta}.$$

By substituting from (6.14) into the first of these

$$\partial_\zeta f f^{-1} + \sum_0^N \frac{f \theta_\alpha f^{-1}}{a_\alpha - \zeta} = \partial_\zeta \tilde{f} \tilde{f}^{-1} + \sum_0^N \frac{\tilde{f} \tilde{\theta}_\alpha \tilde{f}^{-1}}{a_\alpha - \zeta} \, .$$

Both sides must therefore be equal to a global rational function on ζ, with simple poles at a_α. We denote this by

$$A = \sum_0^N \frac{A_\alpha}{\zeta - a_\alpha} \, .$$

Since $\sum \mathcal{L}_\alpha = 0$, we also have $\sum A_\alpha = 0$, so the corresponding Fuchsian equation has no pole at infinity: it is the same Fuchsian equation as the one we constructed above by a more abstract argument.

Now from the construction, we see that

$$\frac{\partial f}{\partial \zeta} = A f, \qquad \frac{\partial f}{\partial a_\alpha} = -\frac{A_\alpha f}{\zeta - a_\alpha} \, .$$

The compatibility condition for these linear equations is the Schlesinger equation:

$$\frac{\partial A_\alpha}{\partial a_\beta} = \frac{[A_\alpha, A_\beta]}{a_\alpha - a_\beta} \qquad \alpha \neq \beta \, ,$$

which is well known to determine the isomonodromic deformations of a Fuchsian system.

Almost exactly the same argument works if E is given by a family of local trivializations with patching matrices F_{ij}.

6.6.3 The inverse construction

We see that an equivariant bundle over U generates a family of Fuchsian equations all with the same monodromy; the equations in the family are labelled by the positions of the poles a_α, and the dependence of the coefficients A_α is determined by the Schlesinger equations.

Suppose, instead, we start with one Fuchsian equation (6.12), with poles at $\zeta = a_\alpha$, $\alpha = 0, 1, \dots N$; for simplicity, we shall assume that there is no pole at infinity, so $\sum A_\alpha = 0$. Can we construct the corresponding equivariant bundle, and hence the isomonodromic deformations? We shall think of the given equation as being associated with an 'initial line' $Z^\alpha = a_\alpha - \zeta$.

Let y be a invertible matrix solution. Then, of course, y has singularities at the poles and is multi-valued: if we continue it around one of the poles, then it becomes yM for some constant matrix M.

For each α, choose some $\alpha' \neq \alpha$ and put

$$\zeta_\alpha = \frac{a_{\alpha'} Z^\alpha - a_\alpha Z^{\alpha'}}{Z^\alpha - Z^{\alpha'}}.$$

Then $\zeta_\alpha = \zeta$ on the initial line. We also pick an open cover U_α of a neighbourhood U of the initial line in $\mathbb{C}P_N$ so that U_α contains the points $Z^\alpha = 0$, but not any of the points $Z^\beta = 0$ for $\beta \neq \alpha$. We then define a holomorphic vector bundle $E \rightarrow U$ by taking its transition matrices between trivializations over U_α and U_β to be

$$F_{\alpha\beta} = y(\zeta_\alpha) y(\zeta_\beta)^{-1}$$

on $U_\alpha \cap U_\beta$. It does not matter which branch of y we choose since the monodromy matrix M cancels when we replace one branch by another. Clearly, from their form, the Fs satisfy the cocycle condition, and so they determine a holomorphic vector bundle $E \rightarrow U$.

It is claimed that E is equivariant, and that it generates the given equation, together with its isomonodromic deformations. It is not hard to see this, but we look at the argument below only in the special case of P_{VI}, where there are only four poles and it is a little easier to see what is going on.

6.6.4 The sixth Painlevé equation

In the special case that $N = 3$ (four poles) and that E has structure group $\mathrm{SL}(2, \mathbb{C})$, the Schlesinger equation comes down to the sixth Painlevé equation. In this case, there is essentially one independent variable since three of the poles can be moved to $\zeta = 0, 1, \infty$ by a Möbius transformation, and so all that is left to vary is the fourth (or, more invariantly, we can take the independent variable to be the cross-ratio of the four poles).

The case $N = 3$ is also the Yang–Mills case. Thus there we see directly that that the reduction of the Yang–Mills equations by the diagonal subgroup of the conformal group is the sixth Painlevé equation.

To be more explicit, the generators of the diagonal group on spacetime are the three conformal Killing vector fields

$$X' = -z\partial_z - w\partial_w, \qquad Y' = -\tilde{z}\partial_{\tilde{z}} - \tilde{w}\partial_{\tilde{w}}, \qquad Z' = z\partial_z + \tilde{w}\partial_{\tilde{w}}$$

(corresponding to the three vector fields

$$X = \zeta\partial_\zeta, \qquad Y = -\lambda\partial_\lambda - \mu\partial_\mu - \zeta\partial_\zeta, \qquad Z = \mu\partial_\mu$$

on $\mathbb{C}P_3$). If we put $p = -\log w$, $q = -\log \tilde{z}$, $r = \log(\tilde{w}/\tilde{z})$, $t = z\tilde{z}/w\tilde{w}$, then these become

$$X = \partial_p, \qquad Y = \partial_q, \qquad Z = \partial_r.$$

We can choose the gauge for the Yang–Mills field so that

$$\Phi = P\,dp + Q\,dq + R\,dr \tag{6.15}$$

with P constant and Q, R depending on t; then the ASD Yang–Mills equations reduce to

$$tQ' = [R, Q], \qquad t(1 - t)R' = [tP + Q, R].$$

It is straightforward to show that if we take a root x of the quadratic

$$\det[P, xQ - R] = 0,$$

and write $x = (y - t)/y(t - 1)$, then y satisfies the sixth Painlevé equation

$$
\begin{aligned}
y'' &= \frac{1}{2}\left(\frac{1}{y} + \frac{1}{y-1} + \frac{1}{y-t}\right)y'^2 - \left(\frac{1}{t} + \frac{1}{t-1} + \frac{1}{y-t}\right)y' \\
&\quad + \frac{y(y-1)(y-t)}{t^2(t-1)^2}\left(\alpha + \frac{\beta t}{y^2} + \frac{\gamma(t-1)}{(y-1)^2} + \frac{\delta t(t-1)}{(y-t)^2}\right),
\end{aligned}
$$

where α, β and γ are constants (that determine invariants of P, Q, and R). Conversely any solution y determines a self-dual Yang–Mills field of the form (6.15).

6.6.5 Twistor construction of solutions

Let us return to the initial Fuchsian equation. In the case of the sixth Painlevé equation, this is of the form

$$\frac{dy}{d\zeta} = \left(\frac{A_0}{\zeta} + \frac{A_1}{1+\zeta} + \frac{A_t}{\zeta+t}\right)y$$

where we think of ζ as a coordinate on the line $w = z = \tilde{w} = 1$, $\tilde{z} = t$.

We pick an initial value of t, and put $\xi = \zeta/(\mu - \zeta)$, $\tilde{\xi} = \lambda - t_0$ (these are local holomorphic function on $\mathbb{C}P_3$ with the property that $\xi = \tilde{\xi} = \zeta$ on the initial line). Then ξ is constant along Y, W, and $\tilde{\xi}$ is constant along X, Z, while

$$X(\xi) = \frac{\mu\zeta}{(\mu - \zeta)^2}, \qquad Y(\tilde{\xi}) = -\lambda.$$

Therefore if y is a solution of the initial Fuchsian equation

$$X(y(\xi)) = \frac{1}{\mu - \zeta}\left(\mu A_0 + \zeta A_1 + \frac{A_t}{\zeta(1 - t_0) + t_0\mu}\right)y$$

$$Y(y(\tilde{\xi})) = \left(-\lambda\frac{A_0}{\lambda - t_0} + \frac{-\lambda A_1}{\lambda + 1 - t_0} + A_t\right)y.$$

We follow the general template for the Schlesinger equation, and put

$$F = y(\tilde{\xi})y(\xi)^{-1},$$

and take F to be the transition matrix of a holomorphic bundle between two open sets: U containing $\zeta = 0, -1$ and $\tilde{U}$ containing $\zeta = -t_0, \infty$. Then because the expressions in brackets above are holomorphic in U and $\tilde{U}$, respectively, at least near the initial line, we have an equivariant bundle.

If we take some other line given by $w = z = \tilde{w} = 1$, $\tilde{z} = t$, then the extraction of the deformed Fuchsian equation comes down to solving the Riemann–Hilbert problem

$$y(\zeta)y(\zeta + t - t_0)^{-1} = f(t, \zeta)^{-1}\tilde{f}(t, \zeta)$$

with $f, \tilde{f}$ holomorphic in U and $\tilde{U}$ respectively.

6.6.6 The Painlevé test

This construction fits into general pattern: to obtain a family of ODEs from an a holomorphic vector bundle $E \to M$, we need two ingredients. First, an equivariance condition under a Lie algebra of holomorphic vector fields of the same dimension as M, with the vector fields spanning the tangent space at almost every point of M. Second, a family of embeddings $\mathbb{C}P_1 \to M$. The ODE will not generally be Fuchsian, but A will always be rational. The deformations are always isomonodromic.

This fact explains why symmetry reductions to ODEs of the integrable systems that arise from twistor constructions lead to Painlevé and more general isomonodromy deformation equations – in short, why they pass the Painlevé test.

6.6.7 Exercise

(6.1) Show that if instead of X, Y, Z, we take the generators of the subgroup of $\mathrm{PGL}(4, \mathbb{C})$ of matrices of the form

$$
\begin{pmatrix}
a & b & c & d \\
0 & a & b & c \\
0 & 0 & a & b \\
0 & 0 & 0 & a
\end{pmatrix}
$$

then the result is a family of ODEs in which A has a single pole of order 4.

References

Ablowitz, M. J. and Clarkson, P. A. (1991). *Solitons, nonlinear evolution equations and inverse scattering*. London Mathematical Society Lecture Notes in Mathematics, **149**, Cambridge University Press, Cambridge.

Atiyah, M. F. (1979). *Geometry of Yang–Mills fields*. Lezioni Fermiane. Accademia Nazionale dei Lincei and Scuola Normale Superiore, Pisa.

Atiyah, M. F., Hitchin, N. J. and Singer, I. M. (1978*a*). Self-duality in four-dimensional Riemannian geometry. *Proc. Roy. Soc. Lond.*, **A 362**, 425–61.

Atiyah, M. F., Hitchin, N. J., Drinfeld, V. G. and Manin, Yu.I. (1978*b*). Construction of instantons. *Phys. Lett.*, **A65**, 185–7.

Atiyah, M. F. and Ward, R. S. (1977). Instantons and algebraic geometry. *Commun. Math. Phys.*, **55**, 111–24.

Drinfeld, V. G. and Sokolov, V. V. (1985). Lie algebras and equations of Korteweg–de Vries type. *J. Sov. Math.*, **30**, 1975–2036.

Gunning, R. C. (1966). *Lectures on Riemann surfaces*. Princeton Mathematical Notes. Princeton University Press, Princeton, New Jersey.

Hitchin, N. J. (1995). Twistor spaces, Einstein metrics and isomonodromic deformations. *J. Diff. Geom.*, **42**, 30–112.

Kobayashi, S. and Nomizu, K. (1969). *Foundations of differential geometry*, Vol. 2. Wiley, New York.

Mason, L. J. and Sparling, G. A. J. (1989). Nonlinear Schrödinger and Korteweg–de Vries are reductions of self-dual Yang–Mills. *Phys. Lett.*, **A137**, 29–33.

Mason, L. J. and Woodhouse, N. M. J. (1996). *Integrability, self-duality, and twistor theory.* Oxford University Press, Oxford.

Penrose, R. (1976). Nonlinear gravitons and curved twistor theory. *Gen. Rel. Grav.*, **7**, 31–52.

Pressley, A. and Segal, G. B. (1986). *Loop groups.* Oxford University Press, Oxford.

Ward, R. S. (1977). On self-dual gauge fields. *Phys. Lett.*, **61A**, 81–2.

Ward, R. S. (1985). Integrable and solvable systems and relations among them. *Phil. Trans. R. Soc.*, **A315**, 451–7.

Ward, R. S. and Wells, R. O. (1990). *Twistor geometry and field theory.* Cambridge University Press, Cambridge.

7

Transformations and reductions of integrable nonlinear equations and the $\bar{\partial}$-problem

Paolo Maria Santini

Dipartimento di Fisica, Università di Roma "La Sapienza"
Piazz.le Aldo Moro 2, I-00185 Roma, Italy
Istituto Nazionale di Fisica Nucleare, Sezione di Roma
P.le Aldo Moro 2, I-00185 Roma, Italy
paolo.santini@roma1.infn.it

Abstract

The $\bar{\partial}$ dressing method is used to construct and solve integrable nonlinear equations as well as to describe their transformations and reductions. The theory is illustrated on a distinguished example: the quadrilateral lattice and its continuous limit: the conjugate net.

7.1 The $\bar{\partial}$ dressing method

All the nonlinear equations integrable via the Inverse Spectral Transform (ISM) are the compatibility condition of two (or more) linear problems and are characterized by the property that they are linearized in a suitable spectral space. The ISM establishes the proper connection between the configuration space and this spectral space. In the spectral space the underlying mathematical structure is an analyticity problem: a Riemann–Hilbert or, more generally, a $\bar{\partial}$ problem on the complex plane $\mathcal{C}$ [1], [2], [3], [4], [5].

The $\bar{\partial}$ dressing method is a powerful evolution of the ISM; its starting point is not the nonlinear equation and its associated linear systems, but a linear $\bar{\partial}$ problem. A linear (simple) dependence of the $\bar{\partial}$ data on the coordinates implies algebrically that the solution of the $\bar{\partial}$ problem solves a set of linear equations in configuration space, whose integrability condition is the integrable nonlinear equation. Therefore the $\bar{\partial}$ dressing method allows one to construct, at the same time, integrable nonlinear equations together with a large class of their solutions [6], [7], [8]. As we shall see in the following, the $\bar{\partial}$ dressing method provides also the convenient setting in which to study symmetry transformations and symmetry constraints of the integrable nonlinear equation.

These short notes are organized as follows. In Section 7.1.1 we present a short introduction to the theory of the $\bar{\partial}$ problem. In Section 7.1.2 we discuss the basic ideas of the $\bar{\partial}$ dressing method and in Section 7.2 we illustrate the method on a basic example: the quadrilateral lattice and its continuous limit: the conjugate net. In Section 7.3.1 we briefly discuss the basic symmetry transformations in the $\bar{\partial}$ formalism and we derive the basic bilinear formulas which are used in Section 7.3.2 to construct a class of symmetry constraints leading to basic geometric reductions of the quadrilateral lattice and of the conjugate net. These explicit reductions are considered in Sections 7.3.3–5.

7.1.1 The $\bar{\partial}$ problem

We first give a short introduction to the $\bar{\partial}$ (DBAR) problem; more details can be found in [5] and [4].

It is well known that, is $\phi(\lambda, \bar{\lambda})$ is analytic in the domain $\mathcal{D}$, then $\partial_{\bar{\lambda}}\phi = 0$, $\lambda \in \mathcal{D}$, where $\partial_{\bar{\lambda}} = \partial/\partial\bar{\lambda}$, and the famous Cauchy formulas hold:

$$\int_{\partial\mathcal{D}} \phi(\lambda)d\lambda = 0, \qquad (7.1)$$

$$\phi(\lambda) = \frac{1}{2\pi i} \int_{\partial\mathcal{D}} \frac{\phi(\lambda')}{\lambda' - \lambda}d\lambda', \quad \lambda \in \mathcal{D}. \qquad (7.2)$$

Consider now a function $\phi(\lambda, \bar{\lambda})$ which is not analytic in $\mathcal{D}$, whose departure from analyticity is described by the equation

$$\partial_{\bar{\lambda}}\phi(\lambda, \bar{\lambda}) = h(\lambda, \bar{\lambda}), \quad \lambda \in \mathcal{D} \qquad (7.3)$$

Our goal is to generalize the Cauchy equations (7.1) and (7.2) to this more general situation. The basic tool is provided by the well-known Gauss–Green formula

$$\int_{\partial A} (Pdy - Qdx) = \int_A (\partial_x P + \partial_y Q)dx \wedge dy, \tag{7.4}$$

where A is a simply connected domain of the (x, y) plane, and by its special cases

$$-\int_{\partial A} gdx = \int_A \partial_y gdx \wedge dy, \tag{7.5}$$

$$\int_{\partial A} gdy = \int_A \partial_x gdx \wedge dy, \tag{7.6}$$

obtained setting $P = 0, Q = g(x, y)$ and $P = g(x, y), Q = 0$ respectively. Through the usual change of variables $(x, y) \to (\lambda, \bar{\lambda})$:

$$\lambda = x + iy, \quad \bar{\lambda} = x - iy, \tag{7.7}$$

of the plane, which implies

$$\partial_\lambda = \frac{1}{2}(\partial_x - i\partial_y), \quad \partial_{\bar{\lambda}} = \frac{1}{2}(\partial_x + i\partial_y), \tag{7.8}$$

$$d\lambda \wedge d\bar{\lambda} = -2idx \wedge dy, \tag{7.9}$$

equations (7.5) and (7.6) take the form:

$$\int_{\partial D} f(\lambda, \bar{\lambda})d\lambda = -\int_D \partial_{\bar{\lambda}} f(\lambda, \bar{\lambda})d\lambda \wedge d\bar{\lambda}, \tag{7.10}$$

$$\int_{\partial D} f(\lambda, \bar{\lambda})d\bar{\lambda} = \int_D \partial_\lambda f(\lambda, \bar{\lambda})d\lambda \wedge d\bar{\lambda}, \tag{7.11}$$

where $A \to D$ and $f(\lambda, \bar{\lambda}) = g(x, y)$. Notice that equation (7.10) provides the wanted generalization of the Cauchy theorem (7.1); to get the generlization of the Cauchy formula (7.2), we first prove that $1/\pi\lambda$ is the localized Green's function of the $\partial_{\bar{\lambda}}$ operator; i.e.:

$$\partial_{\bar{\lambda}}\left(\frac{1}{\lambda - \lambda_0}\right) = \pi\delta(\lambda - \lambda_0), \tag{7.12}$$

where the delta function is defined in the natural way

$$\delta(\lambda - \lambda_0) = \delta(\lambda_R - \lambda_{0R})\delta(\lambda_I - \lambda_{0I}) \tag{7.13}$$

$(\lambda_R = Re\lambda, \lambda_I = Im\lambda)$; which implies, from (7.9), that

$$\int_D \delta(\lambda - \lambda_0)f(\lambda, \bar{\lambda})d\lambda \wedge \bar{\lambda} = -2if(\lambda_0, \bar{\lambda}_0), \quad \lambda_0 \in D. \tag{7.14}$$

138

Therefore the formal inverse of the $\partial_{\bar{\lambda}}$ operator is given by

$$\partial_{\bar{\lambda}}^{-1} = \frac{1}{2\pi i} \int \frac{d\lambda' \wedge d\bar{\lambda}'}{\lambda' - \lambda}. \tag{7.15}$$

To prove equation (7.12) we choose $f = \frac{a(\lambda)}{\lambda - \lambda_0}$ in (7.10), where $a(\lambda)$ is analytic in $\mathcal{D}$, and we obtain, using the Cauchy theorem (7.1),

$$2\pi i a(\lambda) = -\int_{\mathcal{D}} a(\lambda) \partial_{\bar{\lambda}} (\frac{1}{\lambda - \lambda_0}) d\lambda \wedge d\bar{\lambda}, \tag{7.16}$$

valid $\forall a(\lambda)$; which implies equation (7.12).

If we choose instead $f = \frac{\phi(\lambda, \bar{\lambda})}{\lambda - \lambda_0}$ in (7.10) and we use (7.12), we obtain the following generalization of the Cauchy formula (7.2):

$$\phi(\lambda, \bar{\lambda}) = \frac{1}{2\pi i} \int_{\partial \mathcal{D}} \frac{\phi(\lambda', \bar{\lambda}')}{\lambda' - \lambda} d\lambda' + \frac{1}{2\pi i} \int_{\mathcal{D}} \frac{d\lambda' \wedge d\bar{\lambda}'}{\lambda' - \lambda} \partial_{\bar{\lambda}} \phi(\lambda', \bar{\lambda}'), \quad \lambda \in \mathcal{D}. \tag{7.17}$$

We conclude these general considerations remarking that, while the $\bar{\partial}$ equation (7.3) admits the general solution

$$\phi(\lambda, \bar{\lambda}) = a(\lambda) + \frac{1}{2\pi i} \int_{\mathcal{D}} \frac{d\lambda' \wedge d\bar{\lambda}'}{\lambda' - \lambda} h(\lambda', \bar{\lambda}'), \quad \lambda \in \mathcal{D}, \tag{7.18}$$

where $a(\lambda)$ is an arbitrary analytic function in $\mathcal{D}$, the $\bar{\partial}$-boundary value problem

$$\partial_{\bar{\lambda}} \phi(\lambda, \bar{\lambda}) = h(\lambda, \bar{\lambda}), \quad \lambda \in \mathcal{D} \tag{7.19}$$

$$\phi(\lambda, \bar{\lambda}) = B(\lambda, \bar{\lambda}), \quad \lambda \in \partial \mathcal{D} \tag{7.20}$$

is solvable only if the boundary condition B and the forcing h satisfy the equation

$$B(\lambda, \bar{\lambda}) = \frac{1}{2\pi i} \int_{\partial \mathcal{D}} \frac{B(\lambda', \bar{\lambda}')}{\lambda' - \lambda} d\lambda' + \frac{1}{2\pi i} \int_{\mathcal{D}} \frac{d\lambda' \wedge d\bar{\lambda}'}{\lambda' - \lambda} h(\lambda', \bar{\lambda}'), \quad \lambda \in \partial \mathcal{D}, \tag{7.21}$$

which follows directly from equation (7.17) in the limit $\lambda \to \zeta \in \partial \mathcal{D}$.

The two solvable $\bar{\partial}$-boundary value (BV) problems which are relevant in the $\bar{\partial}$ dressing are matrix $M \times M$ $\bar{\partial}$-problems and take both the following form:

$$\partial_{\bar{\lambda}} \phi(\lambda, \bar{\lambda}) = \partial_{\bar{\lambda}} \eta(\lambda) + \int_{\mathcal{D}} \phi(\lambda') R(\lambda', \bar{\lambda}', \lambda, \bar{\lambda}) d\lambda' \wedge d\bar{\lambda}', \quad \lambda \in \mathcal{D}, \tag{7.22}$$

$$\phi(\lambda, \bar{\lambda}) = B(\lambda, \bar{\lambda}), \quad \lambda \in \partial \mathcal{D}, \tag{7.23}$$

where $\eta(\lambda)$ is a given normalization of $\phi(\lambda, \bar{\lambda})$ and $R(\lambda', \bar{\lambda}', \lambda, \bar{\lambda})$ is a given $M \times M$ matrix $\bar{\partial}$-datum.

In the first case, the so-called 'canonical' $\bar{\partial}$-BV problem, $\mathcal{D} = \mathcal{C}$ and $\phi \to I$, $\lambda \to \infty$; therefore $\eta = 1$. We call hereafter the solution of the canonical $\bar{\partial}$ BV problem $\chi(\lambda)$, omitting for simplicity its dependence on $\bar{\lambda}$.

In the second case, the 'simple pole normalization' case, $\mathcal{D} = \mathcal{C}_\mu$, where $\mathcal{C}_\mu$ is the complex plane without a small circle around $\lambda = \mu$ and the boundary condition reads:

$$\phi \to 0, \ \lambda \to \infty; \quad \phi \sim \frac{1}{\lambda - \mu}, \quad \lambda \sim \mu. \tag{7.24}$$

Therefore in this case $\eta = (\lambda - \mu)^{-1}$. We call hereafter the solution of this $\bar{\partial}$ BV problem $\chi(\lambda, \mu)$, omitting for simplicity its dependence on $\bar{\lambda}$.

It follows that the solutions of these two $\bar{\partial}$ BV problems are expressed respectively in terms of the following linear integral equations

$$\chi(\lambda) = I + \frac{1}{2\pi i} \int_{\mathcal{C}} \frac{d\lambda' \wedge d\bar{\lambda}'}{\lambda' - \lambda} \int_{\mathcal{C}} \chi(\lambda'') R(\lambda'', \lambda') d\lambda'' \wedge d\bar{\lambda}'' \ \lambda \in \mathcal{C}, \tag{7.25}$$

$$\chi(\lambda, \mu) = \frac{1}{\lambda - \mu} + \frac{1}{2\pi i} \int_{\mathcal{C}_0} \frac{d\lambda' \wedge d\bar{\lambda}'}{\lambda' - \lambda} \int_{\mathcal{C}_0} \chi(\lambda'', \mu) R(\lambda'', \lambda') d\lambda'' \wedge d\bar{\lambda}'',$$
$$\lambda \in \mathcal{C}_0, \tag{7.26}$$

omitting hereafter the dependence of R on $\bar{\lambda}, \bar{\lambda}'$.

In all our considerations we shall assume that the solution of the $\bar{\partial}$ problem (7.22), expressed through the linear integral equation

$$\phi(\lambda) = \eta(\lambda) + \frac{1}{2\pi i} \int_{\mathcal{C}} \frac{d\lambda' \wedge d\bar{\lambda}'}{\lambda' - \lambda} \int_{\mathcal{C}} \phi(\lambda'') R(\lambda'', \lambda') d\lambda'' \wedge d\bar{\lambda}'' \ \lambda, \lambda' \in \mathcal{C},$$
$$\tag{7.27}$$

be unique; this implies that the solution of the homogeneous $\bar{\partial}$ problem satisfying the homogeneous boundary condition $\phi \to 0$, $\lambda \to \infty$, and corresponding therefore to the case $\eta = 0$, is $\phi = 0$.

It is important to introduce also the "adjoint" $\bar{\partial}$ problem:

$$\partial_{\bar{\lambda}} \phi^*(\lambda) = -\partial_{\bar{\lambda}} \eta(\lambda) - \int_{\mathcal{C}} R(\lambda, \lambda') \phi^*(\lambda') d\lambda' \wedge d\bar{\lambda}' \ , \quad \lambda, \lambda' \in \mathcal{C}. \tag{7.28}$$

The $\bar{\partial}$ problem (7.22) and its adjoint (7.28) imply the basic bilinear identity:

$$\int_{\gamma_\infty} \phi_1(\lambda) \phi_2^*(\lambda) d\lambda + \int_{\mathcal{C}} [(\partial_{\bar{\lambda}} \eta_1)(\lambda) \phi_2^*(\lambda) - \phi_1(\lambda) \partial_{\bar{\lambda}} \eta_2(\lambda)] d\lambda \wedge d\bar{\lambda} +$$

$$\int_C \int_C \phi_1(\lambda')[R_1(\lambda', \lambda) - R_2(\lambda', \lambda)]\phi_2^*(\lambda)d\lambda \wedge d\bar{\lambda} \ d\lambda' \wedge d\bar{\lambda}' = 0, \quad (7.29)$$

where γ_∞ is the circle with center at the origin and arbitrarily large radius (and the corresponding integration is counter-clockwise). Equation (7.29) involves the solutions ϕ_1 of (7.22), corresponding to the normalization η_1 and to the $\bar{\partial}$ datum R_1, and ϕ_2^* of (7.28), corresponding to the normalization η_2 and to the $\bar{\partial}$ datum R_2 respectively. To obtain this identity, multiply equation (7.22) for ϕ_1 from the right by ϕ_2^* and equation (7.28) for ϕ_2^* from the left by ϕ_1; add the resulting equations, apply the operator $\int_C d\lambda \wedge \lambda'$ and finally use the identity

$$\int_C d\lambda \wedge \lambda' \partial_{\bar{\lambda}} f = -\int_{\gamma_\infty} d\lambda f. \quad (7.30)$$

Of particular importance in the following, together with the two basic solutions $\chi(\lambda)$ and $\chi(\lambda, \mu)$ of equations (7.22), corresponding respectively to the 'canonical normalization' $\eta = 1$ and to the 'simple pole normalization' $\eta = (\lambda - \mu)^{-1}$, will be also the solutions $\chi^*(\lambda)$ and $\chi^*(\lambda, \mu)$ of the corresponding adjoint problems (7.28) .

The solutions of the $\bar{\partial}$ problem (7.22) are expressed in terms of the integral equation (7.27). Explicit solutions can be found for special choices of the arbitrary $\bar{\partial}$ kernel R; in particular, if this kernel is separable, the integral equation reduces to an algebraic system of equations (see for instance [4] and [9]).

7.1.2 The $\bar{\partial}$-dressing method

The $\bar{\partial}$-dressing method is a powerful method to construct, starting from the general linear $\bar{\partial}$ problem (7.22) in the spectral space (the λ plane), integrable nonlinear equations in the configuration space parametrized by the, in general complex, parameters $x = (x_1, .., x_N)$, together with large classes of solutions.

The dependence on the set $x = (x_1, .., x_N)$ of (space) parameters is introduced through the basic function $\psi_0(x.\lambda)$ satisfying the following differential equations:

$$\partial_i \psi_0 = K_i(x, \lambda)\psi_0, \quad i = 1, .., N \quad (7.31)$$

where $\partial_i = \partial/\partial x_i$ and/or the following difference equations:

$$T_i \psi_0 = A_i(x, \lambda)\psi_0, \quad i = 1, .., N, \quad (7.32)$$

where T_i is the translation operator in the x_i variable:

$$T_i f(x) = f(x_1, .., x_i + 1, .., x_N). \tag{7.33}$$

The linear equations (7.31) and (7.32) must be compatible; i.e. the matrix functions $K_i(x, \lambda)$, $A_i(x, \lambda)$ must satisfy the equations:

$$\partial_i K_j - \partial_j K_i + [K_j, K_i] = 0, \tag{7.34}$$

$$(T_j A_i) A_j - (T_i A_j) A_i = 0, \tag{7.35}$$

$$(T_j K_i) A_j - \partial_i A_j - A_j K_i = 0. \tag{7.36}$$

The dressing method is based on the assumption that the $\bar{\partial}$ datum R depends on the parameters x through ψ_0 in the following form:

$$R(\lambda', \lambda; x) = \psi_0(x, \lambda') R_0(\lambda', \lambda)(\psi_0(x, \lambda))^{-1} \tag{7.37}$$

where $R_0(\lambda', \lambda)$ is independent of x. This structure of R and equations (7.34), (7.35), (7.36) imply that, if ϕ and ϕ^* are solutions of the $\bar{\partial}$ problem (7.22) and of its adjoint (7.28), then also the 'covariant' derivative and translations of ϕ:

$$\mathcal{D}_i \phi(\lambda) = \partial_i \phi + \phi K_i, \tag{7.38}$$

$$\mathcal{T}_i \phi(\lambda) = (T_i \phi) A_i, \tag{7.39}$$

and their adjoint operations:

$$\mathcal{D}_i^* \phi^*(\lambda) = -\partial_i \phi^* + K_i \phi^*, \tag{7.40}$$

$$\mathcal{T}_i^+ \phi(\lambda) = A_i^{-1}(T_i \phi^*), \tag{7.41}$$

$$\mathcal{T}_i^- \phi(\lambda) = T_i^{-1}(A_i \phi^*) \tag{7.42}$$

($\mathcal{T}_i^+$ and $\mathcal{T}_i^-$ correspond to forward and backward adjoint translations respectively) satisfy the same $\bar{\partial}$ problems but with different normalizations:

$$\partial_{\bar{\lambda}}(\mathcal{D}_i \phi) = (\partial_{\bar{\lambda}} \eta) K_i + \phi \partial_{\bar{\lambda}} K_i + \int_C (\mathcal{D}_i \phi) R \tag{7.43}$$

$$\partial_{\bar{\lambda}}(\mathcal{T}_i \phi) = (\partial_{\bar{\lambda}} \eta) A_i + (T_i \phi) \partial_{\bar{\lambda}} A_i + \int_C (\mathcal{T}_i \phi) R \tag{7.44}$$

$$\partial_{\bar{\lambda}}(\mathcal{D}_i^* \phi^*) = -K_i(\partial_{\bar{\lambda}} \eta) + (\partial_{\bar{\lambda}} K_i) \phi^* - \int_C R(\mathcal{D}_i^* \phi^*) \tag{7.45}$$

$$\partial_{\bar{\lambda}}(\mathcal{T}_i^+ \phi^*) = -A_i^{-1}(\partial_{\bar{\lambda}}\eta) + (\partial_{\bar{\lambda}}A_i^{-1})(\mathcal{T}_i\phi^*) - \int_C R(\mathcal{T}_i^+ \phi^*) \qquad (7.46)$$

$$\partial_{\bar{\lambda}}(\mathcal{T}_i^- \phi^*) = -(T_i^{-1}A_i)(\partial_{\bar{\lambda}}\eta) + T_i^{-1}((\partial_{\bar{\lambda}}A_i)\phi^*) - \int_C R(\mathcal{T}_i^- \phi^*). \qquad (7.47)$$

Also **any** polynomial combination $\mathcal{L}(\mathcal{D}_1,..,\mathcal{D}_N)\phi$ of the operators $\mathcal{D}_i$ with matrix coefficients which are functions of x but not of λ solve the $\bar{\partial}$ problem (7.22), with different normalizations (the same considerations hold for the adjoint operators), forming a ring of operators satisfying the $\bar{\partial}$ problem (7.22).

In this ring of operators a crucial role is played by those operators $\hat{\mathcal{L}}$ such that $\hat{\mathcal{L}}\phi$ satisfy the homogeneous $\bar{\partial}$ problem (7.22), corresponding to $\eta = 0$ (i.e. $\hat{\mathcal{L}}\phi \to 0$, $\lambda \to \infty$); the uniqueness of the $\bar{\partial}$ problem implies indeed that $\hat{\mathcal{L}}\phi = 0$. The operators $\hat{\mathcal{L}}$ form the left ideal of the ring; a basis $\{\hat{\mathcal{L}}_i\}$ of such ideal gives rise to a complete set of spectral problems

$$\hat{\mathcal{L}}_i\phi = 0 \qquad (7.48)$$

associated to the $\bar{\partial}$ problem (7.22) [6]. These matrix differential and/or difference equations are the nontrivial result of the dressing of the simple equations (7.31),(7.32); they are of course compatible by construction and their compatibility is equivalent to a set of (integrable) nonlinear equations, which are therefore constructed and solved simultaneously.

7.2 $\bar{\partial}$-formulation of conjugate nets and quadrilateral lattices

In this section we illustrate an important application of the $\bar{\partial}$-dressing method to the solution of the Darboux equations:

$$\partial_i Q_{jk} = Q_{ji}Q_{ik}, \quad i \neq j \neq k \neq i \qquad (7.49)$$

which characterize N-dimensional conjugate nets (CNs) in $\mathcal{R}^M$, i.e. manifolds parametrized by conjugate coordinates [10], and of their difference analogues, the quadrilateral lattice (QL) equations [11],[12],[13]:

$$\Delta_i Q_{jk} = (T_i Q_{ji})Q_{ik}, \quad \Delta_i = T_i - 1, \quad i \neq j \neq k \neq i, \qquad (7.50)$$

which characterize the quadrilateral (or planar) lattices, N-dimensional lattices in $\mathcal{R}^M$, whose elementary quadrilaterals are planar [12].

To obtain solutions of the Darboux equations (7.49) and of the QL equations (7.50) respectively we choose the basic function ψ_0 in the following two ways [14], [9]:

$$\psi_0(x, \lambda) = \exp(\lambda \sum_{k=1}^{N} x_k P_k) = \mathrm{diag}(e^{\lambda x_1}, .., e^{\lambda x_N}, 1, .., 1) \qquad (7.51)$$

$$\psi_0(x, \lambda) = \prod_{k=1}^{N} [I + (\lambda - 1)P_k]^{x_k} = \mathrm{diag}(\lambda^{x_1}, .., \lambda^{x_N}, 1, .., 1) \qquad (7.52)$$

where P_i is the usual ith projection matrix: $(P_i)_{jk} = \delta_{ij}\delta_{ik}$. Therefore:

$$K_i = \lambda P_i, \qquad A_i = I + (\lambda - 1)P_i \qquad (7.53)$$

We remark that equation (7.51) can be obtained in a straightforward way from equation (7.52) through the substitution:

$$\lambda \to e^{\epsilon\lambda} \sim 1 + \epsilon\lambda, \qquad x \to \frac{x}{\epsilon}. \qquad (7.54)$$

Therefore all the relevant formulas for CNs can be immediately derived from the corresponding formulas for QLs; in particular, in spectral space, one has the following simple recipe [9]:

$$f(\lambda) \to f(\lambda); \quad f(1/\lambda) \to f(-\lambda); \quad \lambda f(\lambda) \to f(\lambda).$$

Nevertheless, for the sake of completeness, we shall write down equations for both the CN and the QL.

Let us derive now the complete set of spectral problems associated with the Darboux equations. Let $\chi(\lambda)$ be the canonical solution of the $\bar{\partial}$ problem (7.22):

$$\chi \sim I + \frac{Q}{\lambda}, \quad \lambda \to \infty, \qquad (7.55)$$

then $\mathcal{D}_i\chi$ is also a solution of the homogeneous $\bar{\partial}$ problem with the asymptotics $\mathcal{D}_i\chi \sim \lambda P_i + QP_i$, $\lambda \to \infty$, and so is $P_j\mathcal{D}_i\chi \sim P_jQP_i$, $\lambda \to \infty$. Therefore the combination $\hat{\mathcal{L}}\chi = P_j\mathcal{D}_i\chi - P_jQP_i\chi$ is a solution of the homogeneous $\bar{\partial}$ problem and $\hat{\mathcal{L}}\chi \to 0$, $\lambda \to \infty$. Uniqueness implies

$$\hat{\mathcal{L}}\chi = P_j\mathcal{D}_i\chi - P_jQP_i\chi = 0, \quad i \neq j \qquad (7.56)$$

This is the compatible linear system of spectral problems we were looking for; the Darboux equations (7.49) follow from (7.56) in the $\lambda \to \infty$ limit, using (7.55).

In a similar way one derives the compatible linear system of spectral problems

$$P_j(\mathcal{T}_i - 1)\chi - \mathcal{T}_i(P_jQP_i)\chi = 0, \quad i \neq j \qquad (7.57)$$

which reduces down to the QL equations (7.50) in the $\lambda \to \infty$ limit. Furthermore, if we define the matrix function

$$\psi(\lambda) := \chi(\lambda)\psi_0(\lambda), \tag{7.58}$$

then ψ satisfies

$$\Delta_i \psi_{jk}(\lambda) = (T_i Q_{ji})\psi_{ik}(\lambda), \quad i = 1,..,N, \quad j,k = 1,..,N, \ i \neq j, \tag{7.59}$$

$$[\partial_i \psi_{jk}(\lambda) = Q_{ji}\psi_{ik}(\lambda)], \quad i = 1,..,N, \quad j,k = 1,..,N, \ i \neq j, \tag{7.60}$$

Hereafter the first equation corresponds to the QL case, while the second one, in square brackets, corresponds to the CN and is obtainable either directly or through the straightforward continuous limit (7.54) from the first.

Similar considerations hold for the adjoint equations; if $\chi^*(\lambda)$ is the canonical solution of (7.28) and $\psi^*(\lambda) := (\psi_0(\lambda))^{-1}\chi^*(\lambda)$, then the following adjoint linear systems are satisfied:

$$((T_i^- - 1)\chi^*)(\lambda)P_j = -\chi^*(\lambda)T_i^{-1}(P_i Q P_j), \quad i \neq j, \tag{7.61}$$

$$[(\mathcal{D}_i^* \chi^*)(\lambda)P_j = -\chi^*(\lambda)P_i Q P_j], \quad i \neq j, \tag{7.62}$$

$$\Delta_i \psi_{kj}^*(\lambda) = (T_i \psi_{ki}^*(\lambda))Q_{ij}, \quad i = 1,..,N, \quad j,k = 1,..,N, \ i \neq j, \tag{7.63}$$

$$[\partial_i \psi_{kj}^*(\lambda) = \psi_{ki}^*(\lambda)Q_{ij}], \quad i = 1,..,N, \quad j,k = 1,..,N, \ i \neq j, \tag{7.64}$$

where Q_{ij} are the (ij)-components of the matrix Q.

Also the solution $\chi(\lambda, \mu)$ of the $\bar{\partial}$ problem (7.22), corresponding to the simple pole normalization, plays an important role in the theory. Using exactly the same procedure as before, it is possible to show that the matrix function

$$\psi(\lambda, \mu) := (\psi_0(\mu))^{-1}\chi(\lambda, \mu)\psi_0(\lambda), \tag{7.65}$$

is connected to the canonical solutions of the $\bar{\partial}$ problem through the following equations:

$$\mathcal{D}_i((\psi_0(\mu))^{-1}\chi(\lambda, \mu)) = (T_i \psi^*(\mu))P_i \chi(\lambda), \tag{7.66}$$

$$[\mathcal{D}_i((\psi_0(\mu))^{-1}\chi(\lambda, \mu)) = \psi^*(\mu)P_i \chi(\lambda)],$$

$$\Delta_i \psi_{jk}(\lambda, \mu) = (T_i \psi_{ji}^*(\mu))\psi_{ik}(\lambda), \quad i = 1,..,N, \quad j,k = 1,..,N, \ i \neq j; \tag{7.67}$$

$$[\partial_i \psi_{jk}(\lambda, \mu) = \psi_{ji}^*(\mu)\psi_{ik}(\lambda)], \quad i = 1,..,N, \quad j,k = 1,..,N, \ i \neq j;$$

$$\chi^*(\mu) = \lim_{\lambda \to \infty} \lambda\chi(\lambda,\mu), \quad \chi^*(\lambda) = -\lim_{\mu \to \infty} \mu\chi(\lambda,\mu),$$

$$\chi_{ij}(0,\mu) = -(T_j\chi_{ij}^*(\mu))\chi_{jj}(0), \quad \chi_{ij}(\lambda,0) = \chi_{ii}(0)T_i^{-1}\chi_{ij}(\lambda)).$$

Furthermore, the matrix function

$$\psi^*(\lambda,\mu) := (\psi_0(\mathbf{n},\lambda))^{-1}\chi^*(\lambda,\mu)\psi_0(\mathbf{n};\mu) \qquad (7.68)$$

is connected to $\psi(\lambda,\mu)$ through

$$\psi^*(\lambda,\mu) = \psi(\mu,\lambda), \quad j,k = 1,..,M. \qquad (7.69)$$

From the solutions $\psi(\lambda,\mu)$, $\psi(\lambda)$ and $\psi^*(\lambda)$ of the $\bar{\partial}$ problem one can construct a family of parallel N-dimensional quadrilateral lattices $\{\mathbf{r}_{(\mathbf{k})}\}$, $\mathbf{k} = 1,..\mathbf{M}$ in $\mathcal{R}^M$, together with the corresponding tangent vectors $\mathbf{X}_i$ and Lamé coefficients H_i, through the following formulas:

$$\Omega = \int_C d\lambda \wedge d\bar{\lambda} \int_C d\mu \wedge d\bar{\mu} M^*(\mu)\psi(\lambda,\mu)M(\lambda), \qquad (7.70)$$

$$\mathbf{X}_i = \int_C d\lambda \wedge d\bar{\lambda}\psi_i(\lambda)M(\lambda) = (H_{i(1)}^*, H_{i(2)}^*, .., H_{i(M)}^*), \qquad (7.71)$$

$$\mathbf{X}_i^* = \int_C d\mu \wedge d\bar{\mu} M^*(\mu)\psi_i^*(\mu) = (H_{(1)i}, H_{(2)i}, .., H_{(M)i}), \qquad (7.72)$$

in such a way that

$$\Delta_i \mathbf{r}_{(\mathbf{k})} = (T_i H_{(k)i})\mathbf{X}_i,$$

$$[\partial_i \mathbf{r}_{(\mathbf{k})} = H_{(k)i}\mathbf{X}_i],$$

where $\mathbf{r}_{(\mathbf{k})}$ is the kth row of matrix Ω, $\psi_i(\lambda)$ is the ith row of matrix $\psi(\lambda)$, $\psi_i^*(\mu)$ is the ith column of matrix $\psi^*(\mu)$ and $M(\lambda)$, $M^*(\mu)$ are arbitrary $M \times M$ matrices independent of x.

The evaluation of equations (7.56) at the distinguished point $\lambda = 0$ leads [16] to the spectral formulation of the τ-function of the QL. Indeed, at $\lambda = 0$, equations (7.56) read:

$$\Delta_i \chi_{jj}(0) = (T_i Q_{ji})\chi_{ij}(0),$$

$$\chi_{ji}(0) + (T_i Q_{ji})\chi_{ii}(0) = 0, \qquad (7.73)$$

and imply that

$$\frac{T_i\chi_{jj}(0)}{\chi_{jj}(0)} = 1 - (T_i Q_{ji})(T_j Q_{ij}). \qquad (7.74)$$

Since the RHS of equation (7.74) is symmetric wrt i and j, we can introduce the potential τ, the τ - function of the QL, in the following way:

$$\chi_{ii}(0) = \frac{T_i\tau}{\tau}. \tag{7.75}$$

Using equation (7.75), equation (7.74) becomes

$$\frac{\tau(T_iT_j\tau)}{(T_i\tau)(T_j\tau)} = 1 - (T_iQ_{ji})(T_jQ_{ij}). \tag{7.76}$$

The introduction of the second potential τ allows one to write the τ function representation of the QL equations (7.50) [15]:

$$\tau T_i\tau_{jk} - (T_i\tau)\tau_{jk} = (T_i\tau_{ji})\tau_{ik}, \quad i \neq j \neq k \neq i,$$

$$\tau T_iT_j\tau - (T_i\tau)(T_j\tau) + (T_i\tau_{ji})(T_j\tau_{ij}) = 0 \quad i \neq j,$$

where

$$\tau_{ij} = \tau Q_{ij}, \quad i \neq j.$$

Finally equation (7.73) gives:

$$\chi_{ij}(0) = -\frac{T_j\tau_{ij}}{\tau}, \quad i \neq j. \tag{7.77}$$

and, analogously:

$$\chi_{ii}^*(0) = \frac{T_i^{-1}\tau}{\tau}, \quad \chi_{ij}^*(0) = \frac{T_i^{-1}\tau_{ij}}{\tau}. \tag{7.78}$$

7.3 Transformations and reductions of conjugate nets and quadrilateral lattices

The theory of the symmetry transformations and of the symmetry constraints for integrable nonlinear systems takes a particularly simple form in the $\bar{\partial}$ dressing context. Hereafter we assume, without loss of generality, that the basic matrix function ψ_0 be diagonal.

7.3.1 Symmetry transformations

If $R(\lambda', \lambda)$ is the $\bar{\partial}$ datum associated with a certain nonlinear systems, then equation

$$\tilde{R}(\lambda', \lambda) = \beta(\lambda', \lambda)F(\lambda')R(\lambda', \lambda)F^{-1}(\lambda), \tag{7.79}$$

is a general symmetry transformation of that integrable nonlinear system, where $\beta(\lambda', \lambda)$ is an arbitrary scalar function of λ and λ', and $F(\lambda)$ is an arbitrary diagonal matrix function of λ (but independent of x). We call this symmetry, which always exists, of the 'first kind'.

The particular form of ψ_0 as function of λ often implies the existence of additional symmetries. For instance, for QLs and CNs we have the obvious symmetries:

$$\psi_0(\lambda^{-1}) = \psi_0^{-1}(\lambda),$$

$$[\psi_0(-\lambda) = \psi_0^{-1}(\lambda)], \qquad (7.80)$$

implying that, if $R(\lambda', \lambda)$ is the $\bar{\partial}$ kernel of a QL [CN], then

$$\hat{R}(\lambda', \lambda) = R^T(\lambda^{-1}, \lambda'^{-1}) = \psi_0(\lambda') R_0^T(\lambda^{-1}, \lambda'^{-1}) \psi_0^{-1}(\lambda)$$

$$[\hat{R}(\lambda', \lambda) = R^T(-\lambda, -\lambda') = \psi_0(\lambda') R_0^T(-\lambda, -\lambda') \psi_0^{-1}(\lambda)] \qquad (7.81)$$

are also symmetry transformations [17]. Therefore the symmetry (7.80) implies the existence of the additional symmetry transformation

$$\hat{R}(\lambda', \lambda) = \beta(\lambda', \lambda) F(\lambda') R^T(\lambda^{-1}, \lambda'^{-1}) F^{-1}(\lambda)$$

$$[\hat{R}(\lambda', \lambda) = \alpha(\lambda', \lambda) F(\lambda') R^T(-\lambda, -\lambda') F^{-1}(\lambda)], \qquad (7.82)$$

which we call of 'second kind'.

It is interesting to remark that the 'square' of a second kind symmetry transformation is a symmetry transformation of first kind; indeed:

$$\hat{\hat{R}}(\lambda', \lambda) = \hat{\beta}(\lambda, \lambda') \hat{F}(\lambda') R(\lambda', \lambda) \hat{F}^{-1}(\lambda),$$

$$\hat{\beta}(\lambda', \lambda) = \beta(\lambda', \lambda) \beta(\lambda^{-1}, \lambda'^{-1}), \quad \hat{F}(\lambda) = F(\lambda) F^{-1}(\lambda^{-1}),$$

$$[\hat{\beta}(\lambda', \lambda) = \beta(\lambda', \lambda) \beta(-\lambda, -\lambda'), \quad \hat{F}(\lambda) = F(\lambda) F^{-1}(-\lambda)]; \qquad (7.83)$$

but it is not true that any symmetry transformation of first kind is the square of a symmetry transformation of second kind.

It is possible to show that one of the implications of the symmetry transformation of the second kind (7.82) is that $\lambda^{-2} F(\lambda^{-1}) \hat{\phi}^T(\lambda^{-1})$ satisfies the adjoint $\bar{\partial}$ problem (7.28), while $\hat{\phi}^{*T}(\lambda^{-1}) F^{-1}(\lambda^{-1})$ satisfies the $\bar{\partial}$ problem (7.22):

$$\partial_{\bar{\lambda}}(\lambda^{-2} F(\lambda^{-1}) \hat{\phi}^T(\lambda^{-1})) = \partial_{\bar{\lambda}}(\lambda^{-2} F(\lambda^{-1})) \hat{\phi}^T(\lambda^{-1})$$

$$+ \lambda^{-2} F(\lambda^{-1}) \partial_{\bar{\lambda}} \eta(\lambda^{-1}) - \int_C R(\lambda, \lambda')(\lambda'^{-2} F(\lambda'^{-1}) \hat{\phi}^T(\lambda'^{-1})) d\lambda' \wedge d\bar{\lambda}',$$

$$(7.84)$$

$$\partial_{\bar{\lambda}}(\phi^{*T}(\lambda^{-1})F^{-1}(\lambda^{-1})) = \hat{\phi}^{*T}(\lambda^{-1})\partial_{\bar{\lambda}}F^{-1}(\lambda^{-1})$$

$$- (\partial_{\bar{\lambda}}\eta(\lambda^{-1}))F^{-1}(\lambda^{-1}) + \int_C (\hat{\phi}^{*T}(\lambda'^{-1})F^{-1}(\lambda'^{-1}))R(\lambda',\lambda)d\lambda' \wedge d\bar{\lambda}',$$

$$(7.85)$$

$$[\lambda^{-1} \to -\lambda]$$

where $\hat{\phi}$ and $\hat{\phi}^*$ are the solutions of the $\bar{\partial}$ problems (7.22) and (7.28) corresponding to $\hat{R}$.

These equations, through the bilinear identity (7.29), imply the following nonlocal quadratic relations [17]:

$$\int_{\gamma_\infty} \phi(\lambda)(\lambda^{-2}F(\lambda^{-1}))\hat{\phi}^T(\lambda^{-1})d\lambda + \int_C [\phi(\lambda)\partial_{\bar{\lambda}}(\lambda^{-2}F(\lambda^{-1}))\hat{\phi}^T(\lambda^{-1})$$

$$+\phi(\lambda)F(\lambda)(\partial_{\bar{\lambda}}\eta(\lambda^{-1})) + (\partial_{\bar{\lambda}}\eta(\lambda))F(\lambda)\hat{\phi}^T(\lambda^{-1})]d\lambda \wedge d\bar{\lambda} = 0, \quad (7.86)$$

$$\int_{\gamma_\infty} \hat{\phi}^{*T}(\lambda^{-1})F^{-1}(\lambda^{-1})\phi^*(\lambda)d\lambda + \int_C [\hat{\phi}^{*T}(\lambda^{-1})(\partial_{\bar{\lambda}}F^{-1}(\lambda^{-1}))\phi^*(\lambda)$$

$$-\hat{\phi}^{*T}(\lambda^{-1})F^{-1}(\lambda^{-1})(\partial_{\bar{\lambda}}\eta(\lambda))$$

$$-(\partial_{\bar{\lambda}}\eta(\lambda^{-1}))F^{-1}(\lambda^{-1})\phi^*(\lambda)]d\lambda \wedge d\bar{\lambda} = 0, \quad (7.87)$$

$$[\lambda^{-1} \to -\lambda]. \quad (7.88)$$

Therefore the symmetry transformation of second kind establishes a nontrivial quadratic connection, whose nature depends on the particular choice of $F(\lambda)$, between the solutions of the $\bar{\partial}$ problem (7.22) for R and of the adjoint $\bar{\partial}$ problem (7.28) for $\hat{R}$ or, equivalently, between QL's and the dual objects, the quadrilateral hyperplane lattices [16], of the transformed QL.

7.3.2 Symmetry constraints

The transformation of second kind (7.82) allows to introduce in a natural way the symmetry constraint $\hat{R}(\lambda',\lambda) = R(\lambda',\lambda)$. This constraint can be written down in terms of R in the following way:

$$R(\lambda',\lambda) = |\lambda|^{-4}\bar{\lambda}'^{-2}F(\lambda')R^T(\lambda^{-1},\lambda'^{-1})F^{-1}(\lambda), \quad (7.89)$$

$$[R(\lambda',\lambda) = F(\lambda')R^T(-\lambda,-\lambda')F^{-1}(\lambda)]$$

and is admissible iff

$$F(\lambda) = F_{\pm}(\lambda) = \lambda^{-1}(A(\lambda) \pm A(\lambda^{-1})), \tag{7.90}$$

$$[F_{\pm}(\lambda) = A(\lambda) \pm A(-\lambda)], \tag{7.91}$$

where $A(\lambda)$ is an arbitrary diagonal matrix.

The nonlocal quadratic relations (7.86), (7.87) become the following nonlocal quadratic constraints [16]:

$$\int_{\gamma_\infty} \phi(\lambda)(\lambda^{-2}F(\lambda^{-1}))\phi^T(\lambda^{-1})d\lambda + \int_C [\phi(\lambda)\partial_{\bar\lambda}(\lambda^{-2}F(\lambda^{-1}))\phi^T(\lambda^{-1})$$

$$+\phi(\lambda)F(\lambda)(\partial_{\bar\lambda}\eta(\lambda^{-1})) + (\partial_{\bar\lambda}\eta(\lambda))F(\lambda)\phi^T(\lambda^{-1})]d\lambda \wedge d\bar\lambda = 0, \tag{7.92}$$

$$\int_{\gamma_\infty} \hat\phi^{*T}(\lambda^{-1})F^{-1}(\lambda^{-1})\phi^*(\lambda)d\lambda + \int_C [\phi^{*T}(\lambda^{-1})(\partial_{\bar\lambda}F^{-1}(\lambda^{-1}))\phi^*(\lambda)$$

$$- \phi^{*T}(\lambda^{-1})F^{-1}(\lambda^{-1})(\partial_{\bar\lambda}\eta(\lambda))$$

$$- (\partial_{\bar\lambda}\eta(\lambda^{-1}))F^{-1}(\lambda^{-1})\phi^*(\lambda)]d\lambda \wedge d\bar\lambda = 0, \tag{7.93}$$

$$[\lambda^{-1} \to -\lambda].$$

Therefore the symmetry constraint (7.89) establishes a nontrivial quadratic connection, whose nature depends on the particular choice of $F(\lambda)$, between the solutions of the $\bar\partial$ problem (7.22) and of its adjoint (7.28) or, equivalently, between QL's and their dual objects, the quadrilateral hyperplane lattices [16].

In the following we shall identify the matrix functions $A(\lambda)$ which correspond to the symmetric, circular and Egorov lattices [symmetric, orthogonal and Egorov nets].

7.3.3 $\bar\partial$-formulation of the symmetric lattice

It is possible to show that, for the following choice [16]:

$$A(\lambda) = 1/2 \quad \Rightarrow \quad F_+(\lambda) = \lambda^{-1}I, \Rightarrow$$

$$R(\lambda', \lambda) = |\lambda|^{-4}\bar\lambda'^{-2}\lambda\lambda'^{-1}R^T(\lambda^{-1}, \lambda'^{-1})$$

$$[F_+(\lambda) = I \Rightarrow R(\lambda', \lambda) = R^T(-\lambda, -\lambda')] \tag{7.94}$$

the following equations hold:

$$\psi^T(\lambda, \mu) = (\lambda\mu)^{-1}\psi(\mu^{-1}, \lambda^{-1}), \tag{7.95}$$

$$[\psi^T(\lambda,\mu) = \psi(-\mu,-\lambda)], \qquad (7.96)$$

$$\lambda^{-1}\psi_{ij}(\lambda^{-1}) = \frac{T_i \tau}{\tau}\psi_{ji}^*(\lambda), \qquad (7.97)$$

$$[\psi_{ij}(-\lambda) = \psi_{ji}^*(\lambda)], \qquad (7.98)$$

$$\chi^T(0) = \chi(0), \qquad (7.99)$$

$$[Q^T = Q]. \qquad (7.100)$$

Using equations (7.78) and (7.77) it is easy to identify equation (7.99) with the symmetric constraint of a QL [CN] [16]:

$$T_i(\tau Q_{ji}) = T_j(\tau Q_{ij}), \quad i \neq j, \qquad (7.101)$$

$$[Q_{ij} = Q_{ji}],$$

and equation (7.70) allows to construct a parallel system of symmetric quadrilateral lattices, provided that

$$M^*(\lambda) = \lambda|\lambda|^{-4}M^T(\lambda), \qquad (7.102)$$

$$[M^*(\lambda) = M^T(\lambda)]. \qquad (7.103)$$

7.3.4 $\bar{\partial}$ formulation of the circular lattice

The choice [9] [[14],[18]]:

$$A(\lambda) = (\lambda - 1)I \;\Rightarrow\; F_-(\lambda) = \frac{\lambda+1}{\lambda(\lambda-1)}I \;\; [F_-(\lambda) = \lambda^{-1}I] \qquad (7.104)$$

implies the following consequences of the bilinear constraints (7.92) and (7.93):

$$\chi(0) + \chi^T(0) = 2\chi(1)\chi^T(1), \qquad (7.105)$$

$$[\chi(0)\chi^T(0) = I], \qquad (7.106)$$

$$\chi^*(0) + \chi^{*T}(0) = 2\chi^{*T}(-1)\chi^*(-1), \qquad (7.107)$$

$$[\partial_i Q_{ji} + \partial_j Q_{ij} + \sum_{k\neq i,j} Q_{ki}Q_{kj} = 0, \;\; i \neq j], \qquad (7.108)$$

$$\frac{\lambda+1}{\lambda(1-\lambda)}\chi^T(\lambda^{-1},\mu^{-1}) = \frac{\mu(\mu+1)}{1-\mu}\chi(\mu,\lambda) + \chi(1,\lambda)\chi^T(1,\mu^{-1}), \qquad (7.109)$$

$$\frac{\lambda - 1}{\lambda(1 + \lambda)}\chi^T(\mu^{-1}, \lambda^{-1}) = \frac{\mu(\mu - 1)}{1 + \mu}\chi(\lambda, \mu) + \chi^T(\mu^{-1}, -1)\chi(\lambda, -1),$$

$$(7.110)$$

$$4\chi(1, -1)\chi^T(1, -1) = I. \qquad (7.111)$$

Through the identifications:

$$\mathbf{x}_{(i)} = (\psi_{i1}(1, \mu), .., \psi_{iM}(1, \mu)), \qquad (7.112)$$

$$\mathbf{X}_i = (\psi_{i1}(1), .., \psi_{iM}(1)), \quad H_{(n)i} = \psi^*(\mu), \qquad (7.113)$$

$$\mathbf{X}_i^* = (\psi_{1i}^*(-1), .., \psi_{Mi}^*(-1))^T, \qquad (7.114)$$

one obtains:

$$\chi_{ii}(0) = \frac{T_i \tau}{\tau} = |\mathbf{X}_i|^2, \qquad (7.115)$$

and equation (7.105) coincides with the circularity [orthogonality] constraint

$$\mathbf{X}_i \cdot T_i \mathbf{X}_j + \mathbf{X}_j \cdot T_j \mathbf{X}_i = 0, \quad i \neq j, \qquad (7.116)$$

$$[\mathbf{X}_i \cdot \mathbf{X}_j + \mathbf{X}_j \cdot \mathbf{X}_i = 0, \quad i \neq j].$$

We also recognize in equations (7.108) the Lamé equations for orthogonal nets.

7.3.5 $\bar{\partial}$ formulation of the egorov lattice

The Egorov lattice [19][net [20],[10]] arises when the circular and symmetric constraints are simultaneously satisfied; i.e., when

$$\begin{aligned}
|\lambda'|^{-4}\bar{\lambda}^{-2}R^T(\lambda'^{-1}, \lambda^{-1}) &= \lambda R(\lambda, \lambda')\lambda'^{-1} \\
&= \left(\frac{\lambda + 1}{\lambda(\lambda - 1)}\right)^{-1} R(\lambda, \lambda')\frac{\lambda' + 1}{\lambda'(\lambda' - 1)}.
\end{aligned}$$

This implies the equation:

$$\frac{2\lambda(\lambda - \lambda')}{\lambda'(\lambda' - 1)(\lambda + 1)}R(\lambda, \lambda') = 0, \qquad (7.117)$$

which admits the (unique) distributional solution

$$R(\lambda, \lambda') = \frac{i}{2}\delta(\lambda - \lambda')R(\lambda). \qquad (7.118)$$

Therefore the $\bar{\partial}$ formulation of the Egorov lattice is given in terms of the **local** $\bar{\partial}$ problem

$$\partial_{\bar{\lambda}}\phi(\lambda) = \phi(\lambda)R(\lambda), \qquad (7.119)$$

$$R^T(\lambda^{-1}) = |\lambda|^4\bar{\lambda}^2 R(\lambda), \qquad (7.120)$$

$$[R^T(-\lambda) = R(\lambda)].$$

It is straightforward now to verify that this formulation corresponds to the Egorov lattice [net] [16].

Bibliography

V. E. Zakharov, S. V. Manakov, S. P. Novikov and L. P. Pitaevsky. *Theory of solitons. The inverse problem method.* Plenum Press, 1999.

M. J. Ablowitz and H. Segur. *Solitons and the inverse scattering transform.* SIAM, Philadelphia, 1981.

F. Calogero and A. Degasperis. *Spectral transform and solitons: tools to solve and investigate nonlinear evolution equations, vol.1.* North-Holland. Amsterdam, 1982.

B. G. Konopelchenko. *Solitons in multidimension.* World Scientific, Singapore, 1993.

M. J. Ablowitz and P. A. Clarkson. *Solitons, nonlinear evolution equations and inverse scattering.* Cambridge Univ. Press, Cambridge, 1991.

V. E. Zakharov and S. V. Manakov. Construction of multidimensional nonlinear integrable systems and their solution, *Funk. Anal. Appl.* **19** (1985) 89.

L. V. Bogdanov and S. V. Manakov. The nonlocal $\bar{\partial}$-problem and (2+1)-dimensional soliton equations, *J. Phys. A: Math. Gen.* **21** (1988) L537- L544.

V. E. Zakharov. *On the dressing method* in: *Inverse Methods in Action*, ed. P. C. Sabatier, Berlin, Springer, 1990, 602-623.

A. Doliwa, S. V. Manakov and P. M. Santini. $\bar{\partial}$-reductions of the multidimensional quadrilateral lattice: the multidimensional circular lattice, *Comm. Math. Phys.* **196** (1998) 1-18.

G. Darboux. *Lecons sur le systèmes orthogonaux et le coordonnés curvilignes*, 2-ème ed., Gauthier-Villars, Paris, 1910.

L. V. Bogdanov and B. G. Konopelchenko. Lattice and q-difference Darboux–Zakharov–Manakov systems via $\bar{\partial}$-method, *J. Phys. A: Math. Gen.* **28** (1995) L173-L178.

A. Doliwa and P. M. Santini. Multidimensional quadrilateral lattices are integrable, *Phys. Lett. A* **233** (1997) 365-372.

L. V. Bogdanov and B. G. Konopelchenko. Analytic bilinear approach to integrable hierarchies II. Multicomponent KP and 2D Toda lattice hierarchies, *J. Math. Phys.* **39** (1998) 4701-4728.

V. E. Zakharov. Description of the N-orthogonal curvilinear coordinate systems and hamiltonian integrable systems of hidrodynamic type, I: integration of the Lamé equations, *Duke Math. J.* **94** (1998) 103-139.

A. Doliwa, M. Manas, L. Martinez Alonso, E. Medina and P. M. Santini. Charged free fermion, vertex operators and classical transformations of conjugate nets, *J. Phys. A* **32** (1999) 1197-1216.

A. Doliwa and P. M. Santini. The symmetric, d-invariant and Egorov reductions of the quadrilateral lattice, Solv-int/9907012. *J. Phys. and Geom.* (in press).

P. M. Santini. Symmetry transformations and symmetry constraints of the quadrilateral lattice in the $\bar{\partial}$-formalism, in preparation.

V. E. Zakharov and S. V. Manakov. Reduction in systems integrated by the method of the inverse scattering problem, *Dokl. Math.* **57** (1998) 471-474.

W. K. Schief. Talk given at the Workshop: *Nonlinear systems, solitons and geometry*, Oberwolfach, 1997.

L. Bianchi. *Lezioni di geometria differenziale*, 3-a ed. Zanichelli, Bologna, 1924.

Printed in the United States
by Baker & Taylor Publisher Services